LIFE SCIENCES MISCELLANEOUS PUBLICATIONS
ROYAL ONTARIO MUSEUM

JOHN W. REYNOLDS
Illustrated by
DANIEL L. DINDAL

The Earthworms (Lumbricidae and Sparganophilidae) of Ontario

Publication date: 15 June 1977

ISBN 0-88854-191-0
ISSN 0082-5093

Suggested citation: Life Sci. Misc. Pub., R. Ont. Mus.

ROYAL ONTARIO MUSEUM PUBLICATIONS IN LIFE SCIENCES

The Royal Ontario Museum publishes three series in the Life Sciences:

LIFE SCIENCES CONTRIBUTIONS, a numbered series of original scientific publications, including monographic works.

LIFE SCIENCES OCCASIONAL PAPERS, a numbered series of original scientific publications, primarily short and usually of taxonomic significance.

LIFE SCIENCES MISCELLANEOUS PUBLICATIONS, an unnumbered series of publications of varied subject matter and format.

All manuscripts considered for publication are subject to the scrutiny and editorial policies of the Life Sciences Editorial Board, and to review by persons outside the Museum staff who are authorities in the particular field involved.

JOHN W. REYNOLDS is an Assistant Professor in the Department of Forest Resources, University of New Brunswick, Fredericton, New Brunswick.

DANIEL L. DINDAL is a Professor in the Department of Forest Zoology, College of Environmental Science and Forestry, Syracuse, New York.

PRICE: $8.00

100 Queen's Park, Toronto, Canada M5S 2C6

PRINTED AND BOUND IN CANADA BY THE HUNTER ROSE COMPANY

This book is dedicated to Dr. Gordon E. Gates on the occasion of his 80th birthday and 51st year of publication on the Oligochaeta.

Contents

Foreword

"I would not enter on my list of friends,
(Tho' grac'd with polish'd manners and fine sense
Yet wanting sensibility) the man
Who needlessly sets foot upon a worm".

William Cowper, *The Task* (1784)

Nobody could needlessly set foot upon one of the giant earthworms of Australia or Brazil, large specimens of which may attain lengths of 11 feet and weigh up to 1 pound. But to many people earthworms are lowly insignificant creatures whose main utility is to act as bait for catching larger and more edible animals. The Canadian earthworms are indeed represented only by the smaller, more modest, forms and even though to the uninitiated they all seem to be the same there are in fact several species which are not too difficult to distinguish. In this book Dr. Reynolds has assembled, for the first time, all pertinent data, both systematic and biological, on the Canadian earthworm fauna, and with the aid of a key, and the fine illustrations of Dr. Dan Dindal, any naturalist or fisherman should be able to name accurately the specimens that he has at hand.

Earthworms are a significant component of the soil fauna and their beneficial effects on the agricultural properties of soils have been documented since the time of Darwin. Some idea of the extent of their activity can be obtained by reflecting on the fact that something apparently so permanent as the monument of Stonehenge is being buried at the rate of about seven inches per century as a result of the burrowing activities of earthworms. Because of their effects on the soil there can be little doubt that a proper understanding of these creatures is greatly to man's benefit but, as Dr. Reynolds points out, very little is known of their biology in North America. This book should form a valuable basis for further study of these important aspects.

Dr. Faustus, in part I of Goethe's *Faust*, speaks disparagingly of him who "finds his happiness unearthing worms". And an old Chinese aphorism warns "watch the earthworm; miss the eclipse". But anybody who has spent time investigating and observing the smaller and lesser known animals of this planet knows that there is much intellectual satisfaction to be gained from such efforts. The great Victorian naturalist, Thomas Henry Huxley, likened the uninformed naturalist to a person walking through an art gallery in which nine-tenths of the pictures have their faces to the walls. With the aid of this book a few more pictures are now on view.

The Royal Ontario Museum is fortunate to have persuaded Drs. Reynolds and Dindal to collaborate in producing this book, and zoologists, farmers, fishermen, naturalists, and teachers throughout northern North America should have cause to appreciate their labours.

Ian R. Ball
Assistant Curator of
Invertebrate Zoology
Royal Ontario Museum

Acknowledgments

This project was sponsored in part under a Gerald L. Beadel Research Grant from the Tall Timbers Research Station, Tallahassee, Florida. The author wishes to thank Mr. E.V. Komarek, Sr. of Tall Timbers for his support. The author also expresses his gratitude to Mr. Dennis Clarke of Canadian Motor Industries (Toyota Ltd.) for providing a 4-wheel drive Land Cruiser to conduct this study. The author is grateful to Mr. and Mrs. C.W. Reynolds of Islington, Mr. D.W. Reynolds of Mitchell, and Mr. K. Burns of the Biology Department of Lakefield College for providing laboratory space during the field work. The contribution of Mr. C.E. Meadows, Oligochaetology Laboratory, Knoxville to the various laboratory and statistical analyses in this project is appreciated. The author thanks Mrs. W.M. Reynolds of Tall Timbers for constant encouragement and support during the course of the study, for reviewing the manuscript, and for her comments, criticisms, and suggestions. The author would also like to thank Ms. Jennifer Smith of the Royal Ontario Museum for typing the final drafts of the manuscript and he is especially grateful to Dr. Ian R. Ball of the Royal Ontario Museum for the considerable time and effort he devoted to editing the manuscript, which, in the author's opinion, has improved the text. The author is indebted to the following for providing specimens for examination from collections in their care: Dr. I.R. Ball (ROM), Dr. E.L. Bousfield (CNM), Mr. S. Fuller (ANSP), Dr. S.B. Peck (Carleton University), Dr. M.B. Pettibone (USNM), and Mr. D.P. Schwert (University of Waterloo).

The Earthworms (Lumbricidae and Sparganophilidae) of Ontario

Introduction

Earthworms (Annelida, Clitellata, Oligochaeta) are familiar to almost everyone. In North America, they are one of the most popular forms of live bait for fishing (Harman, 1955); gardeners hold them in high esteem as nature's ploughmen (Darwin, 1881); folklore and scientific accounts tell of their medicinal uses (Stephenson, 1930; Reynolds and Reynolds, 1972), and soil inhabiting vertebrates (moles, voles, etc.) store them as a source of food (Plisko, 1961; Skoczeń, 1970). The role of some species in organic matter decomposition and mineral cycling may be important (Bouché, 1972; Edwards and Lofty, 1972), and a great deal has been written concerning earthworm farming (Myers, 1969; Morgan, 1970; Shields, 1971). Biology students the world over study their anatomy (mainly *Lumbricus terrestris*) in great detail (Whitehouse and Grove, 1943). The great amount of literature that has been devoted to a group of organisms that are neither pests nor sources of human nutrition is truly amazing, yet their biology and distribution are still relatively unknown. Many of the world's hundreds of megadrile (= terrestrial oligochaetes) species are known only from a limited series of one or a few specimens.

This text has been designed to introduce the non-specialist to the taxonomy, nomenclature, morphology, distribution, and general biology of earthworms in Ontario and neighbouring areas. The identity, distribution, and habitats of these animals have been surveyed for a variety of habitats in each of the southern counties and districts of the province. An illustrated glossary is included together with a new key to the identification of the earthworms of Ontario that also is applicable to the rest of eastern Canada and to the northern tier of states of the United States. French and English common names are included for each species.

The first records of earthworms from Ontario were provided by Eisen (1874). Recently Reynolds (1972a) reviewed the complete published and verified unpublished records of terrestrial earthworms from this province, and a second report examined those data quantitatively for habitat factors governing megadrile activity in the Haliburton Highlands (Reynolds and Jordan, 1975). This study is a continuation of those reports and presents subsequent collections from 50 counties and districts of southern Ontario in detail (Fig. 1). In addition, unpublished records derived from collections in North American museums and universities, including records from northern Ontario, are presented for the first time.

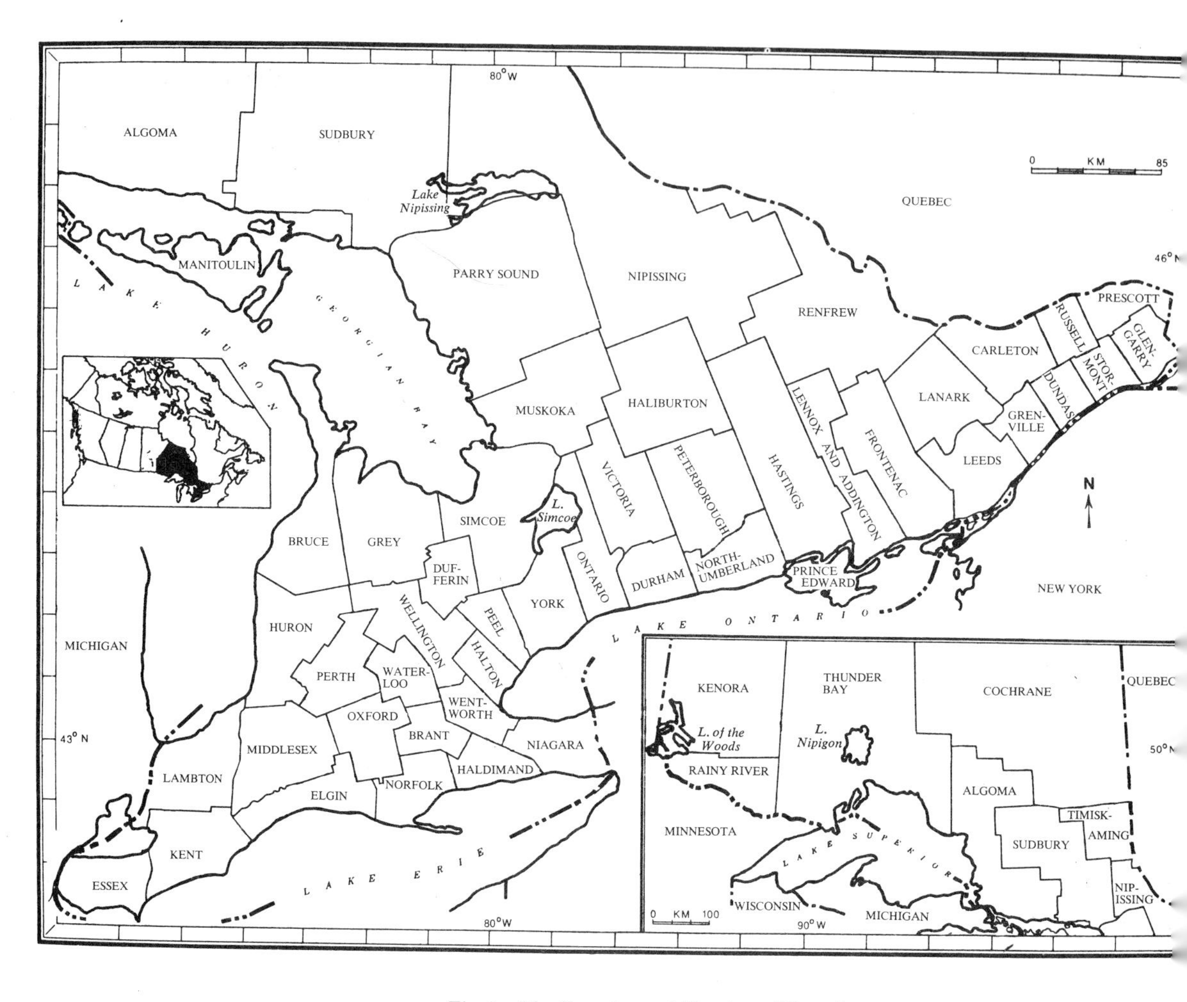

Fig. 1 The Counties and Districts of Ontario.

Thirty-eight of the counties and districts have never had any earthworms reported previously.

At present, there are insufficient megadrile data available to utilize fully some of the habitat information. For example, it would be unwise to try to correlate in detail megadrile distribution with the distribution of soil types until additional surveys from other parts of the continent are completed (cf. Jordan et al., 1976). A preliminary examination of megadrile–vegetation relationships has been presented recently (Reynolds, 1976b). A provincial description is included in an appendix to assemble regional habitat information for future use as well as to familiarize native and foreign readers with Ontario.

The technical terms and conventions necessary for earthworm discussion will be found in the Glossary (p. 18). For additional information on earthworm terminology Stephenson (1930), Causey (1952), Gerard (1946), Ljungström (1970), or Gates (1972c) may be consulted.

General Biology

Introductory Remarks

There can be little doubt that earthworms are the best known of all soil animals. It is common knowledge that they have a beneficial effect on the structure and properties of the soil and that they influence the decomposition processes in organic materials. However, it cannot be denied that much work purporting to demonstrate these aspects has been far from rigorous. In fact, far less is known than is generally believed, and most work is applicable only to Europe. The problem is compounded by the fact that many of the hundreds of described species are known only from a morphological study of a few individuals. Fortunately, nearly all of the species present in Ontario and the neighbouring areas are the widely distributed European species that have received the greatest attention. Major limitations to the interpretation of the literature have been old nomenclatural and taxonomic designations (Reynolds, 1973b).

Sources of information on various biological attributes for species found in Ontario and the surrounding region are Evans and Guild (1948), Bouché (1972), Edwards and Lofty (1972), Gates (1972c), Reynolds (1973d), and Reynolds et al. (1974). Recent reviews of earthworm activity will be found in Kevan (1962) and Wallwork (1970).

General Activity

The main activities of earthworms that affect the soil involve the ingestion of soil and the mixing of the main soil ingredients of clay, lime, and humus; the production of castings of a fine crumb structure which are ejected on the soil surface by some species; the construction of burrows that enhance aeration, drainage, and root penetration; and the production of a tilth that makes suitable habitats for the smaller soil fauna and micro-organisms. It should be remembered, however, that not all Lumbricidae work in the same manner. Some, for example, burrow deeply whereas others do not.

The influence of earthworms on the translocation of soil material may be quite considerable. There have been abundance estimates as high as three million worms per acre and their role in soil fertility is very important. Studying forms that eject casts to the surface, Darwin (1881) estimated that between 7½ and 18 tons of soil per acre per year (about 3 cm per 10 years) can be moved, and the burial of many Roman ruins in Europe has been attributed to the activity of earthworms (Atkinson, 1957).

Earthworms are omnivorous and can utilize many materials in the soil as food, including plant remains, and occasionally animal remains. Lumbricids can withstand considerable starvation and, in *L. terrestris* at least, a water loss of up to 70% of the body weight. Some species can withstand total immersion in water for many weeks, though normally they avoid waterlogged soils.

The reproductive cycle of many Lumbricidae is quite straightforward. Although hermaphrodite, they possess a mechanism to prevent self-fertilization. During copulation the two worms lie side by side with their anterior ends overlapping. A mucous sheath envelops the worms and holds them tightly together. Sperm are released from the testes and flow down the seminal groove in the side of each worm to the spermathecae of its partner. Both worms do this at the same time. Some time after copulation has taken place, and after the worms have separated, the egg cocoons are formed. A mucous tube or belt is secreted around the clitellum. The worm then wriggles out of this belt and as the belt passes the female apertures the eggs are deposited in it. Spermatozoa to fertilize the eggs are deposited as it passes the spermathecal openings. On release the ends of the belt close over to form a cocoon in which the young worms develop.

Cross-fertilization does not occur in all earthworms, however, despite assertions to the contrary in many textbooks. In some species there is parthenogenesis, with concomitant reduction of the male apparatus. Pseudogamy, in which sperm play no part in the development of the egg other than as a stimulant, also may occur. Thus, even if copulation has been observed, the exchange of sperm alone is not evidence for amphimixis. The whole question of reproduction in earthworms has been reviewed by Reynolds (1974c).

Parasites and Predators

Some earthworms (*Allolobophora chlorotica* and *Eisenia rosea*) are parasitized by *Pollenia rudis* (Fabr.), a calliphorid fly known as the cluster fly, which may lay its eggs directly in the earthworm or merely on the surface of the soil (Thomson and Davies, 1973a, 1973b). Cluster flies are the most common and annoying of the flies that overwinter in buildings. Other insects such as ants and beetles are predaceous on earthworms (McLeod, 1954). Furthermore, some earthworms may act as intermediate hosts of parasitic worms that affect domestic animals (Kevan, 1962). Reports of mites (Acari) parasitizing earthworm cocoons and adults (*Allolobophora chlorotica* and *Eiseniella tetraedra*) were made by Stone and Ogles (1953) and Oliver (1962).

Earthworms are also an important component of the diet of many birds and mammals. In Europe moles may store them as a source of food (Skoczeń, 1970; Gates, 1972c), usually after biting off four or five of the anterior segments to prevent the worms from escaping (Evans, 1948b). In North America they are eaten by many organisms, including some of economic or recreational importance. According to Liscinsky (1965), for example, the diet of the woodcock (*Philohela minor* Gmelin), a favourite game bird in eastern North America, is primarily earthworms. From my current surveys, and from gut analyses of woodcock, it appears that in the area bounded by Ontario to Nova Scotia and Minnesota to Maryland, 90% of the earthworms in the diet of these birds are *Aporrectodea tuberculata, Dendrobaena octaedra, Dendrodrilus rubidus*, and *Lumbricus rubellus*. Snakes, too, may prey extensively on earthworms. This is true especially of two of our most common species, the red-bellied snake (*Storeria occipitomaculata occipitomaculata* Say) and the eastern garter snake (*Thamnophis sirtalis sirtalis* Linnaeus), and perhaps of four or five other species as well (Logier, 1958). As this book is in proof, the author has examined the gut contents from *Thamnophis butleri* Cope collected in Essex and Lambton Counties, Ontario. The earthworms identified in these snakes' stomachs were *Allolobophora chlorotica, Aporrectodea tuberculata*, and *Lumbricus terrestris*. According to the author and Dr. S.W. Gorham (pers. comm.), this is the first valid report of earthworm species identified from snake stomachs in North America. A recent account was presented by Dindal (1970) of a terrestrial turbellarian,

Bipalium adventitium Hyman, attacking *Dendrodrilus rubidus* and *Lumbricus terrestris*. This flatworm is currently a major problem in outdoor earthworm beds in central New York state (Dindal, pers. comm.).

Environmental Requirements and the Effects of Pesticides

Daylight and ultraviolet light are injurious to earthworms unless the intensity is very low. Temperature relations have been reviewed by Reynolds (1973a), and Gates (1970) quotes interesting accounts of lumbricids studied from the Arctic circle; *Eisenia foetida*, for example, has been found in snow, even though generally associated with warm habitats such as manure piles, and it remains vigorous below 5° C. In Maine *L. terrestris* has been seen copulating while bathed with melt water, and other individuals crawled from under the ice and remained active (Gates, 1970).

The pH tolerance (see Glossary) of earthworms varies from species to species (Reynolds, 1973d). Usually they occur in soil with a pH range of about 4.5 to 8.7 and the earthworm density diminishes as the soil acidity increases. Generally speaking, the greatest earthworm densities are found in neutral soils.

The type of soil also may influence the distribution and abundance of the various species. Gates (1961), for example, divides the earthworms of Maine into three groups depending upon whether or not they are geophagous, in that they pass much soil through the intestine; limiphagous (mud-eating) or limicolous (mud-inhabiting); or, finally, litter-feeding, and hence found primarily in organic matter. From his studies in Sweden, Julin (1949) divided the Lumbricidae into four ecological groups. These were hemerophiles, species favoured by human culture; hemerophobes, species averse to culture; hemerodiaphores, species indifferent to the influence of culture; and hemerobionts, species entirely dependent on culture. Julin's classification has never been applied to the North American Lumbricidae with the exception of a preliminary attempt for the earthworms of Tennessee by Reynolds et al. (1974). Regrettably, there are as yet insufficient data to permit an attempt for the Ontario earthworms; this is a topic worthy of further study.

The application of pesticides to control soil pests, or the earthworm parasites mentioned above, may also kill the earthworms. This devastating effect on earthworm populations has frequently occurred after the application of orchard sprays. Fruit growers have long held earthworms in high esteem for their help in controlling the disease apple scab which is produced by the fungus *Venturia inequalis* (Cke.) Wint. This disease overwinters on the fallen leaves in the orchard. One method of cultural control is to burn the fallen leaves and twigs in the fall of the year. An equally effective and less costly method is to introduce earthworms (preferably *Lumbricus terrestris*), which will pull the fallen leaves into the soil for food and eventual decomposition. According to the findings of Reynolds and Jordan (1975), for example, earthworms have a distinct preference for apple leaves over those of maple. Once the leaves are beneath the soil surface the conidiospores of the fungus are ineffectual inoculating agents of the disease. The preventive measure most commonly used for control of apple scab is frequent spraying of copper sulphate solutions which are toxic to earthworms (Raw and Lofty, 1959).

Many studies have been conducted to determine the effects of pesticides on earthworms. There is little effect on earthworms with normal doses of Aldrin (Edwards and Dennis, 1960; Edwards et al., 1967; Hopkins and Kirk, 1957; Legg, 1968), or benzene hexachloride (BHC) (Dobson and Lofty, 1956; Lipa, 1958; Morrison, 1950); chlordane is extremely toxic to them (Doane, 1962; Edwards, 1965; Hopkins and Kirk, 1957; Schread, 1952). DDT, of course, has been studied by many workers. In general, the application of this pesticide at normal rates does not harm earthworms (Baker, 1946; Doane, 1962; Edwards, 1965; Edwards and Dennis, 1960; Edwards et al., 1967; Hopkins and Kirk, 1957; Thompson, 1971).

Although earthworms are not susceptible to many pesticides at normal dosages, they do concentrate these toxic chemicals in their tissues. Since many of these chemicals have long-lasting residual periods in the soil, there is ample opportunity for earthworms to absorb them from the soil. The importance of this phenomenon is that these pesticides can become concentrated in the food chain. Earthworms are eaten by many species of birds and certain species of amphibians, reptiles, and mammals, which can continue to concentrate these pesticides in their bodies (Hunt and Sacho, 1969). Additional reports of pesticides and their effects on earthworms can be found in Edwards and Lofty (1972).

Herbicides, another group of chemicals, also can affect earthworm populations (Edwards, 1970; Fox, 1964). These chemicals may kill earthworms directly, or indirectly by killing the vegetation on which they feed.

One last group of potential poisons that could become concentrated in the food chain are metal residues. Recently, Gish and Christensen (1973) found that concentrations of certain metals (cadmium, nickel, lead, and zinc) in earthworms were many times that of the surrounding soils. This study was the first report of metal residues in earthworms. Because of the earthworms' position in the food chain and the current studies in other fields on metal toxicity, this is an area requiring further investigation.

Rearing and Culturing Earthworms

It may be of interest to some readers to discuss briefly the rearing or culturing of earthworms. This is not difficult for the species found in Ontario. There are many books available describing techniques (e.g., Ball and Curry, 1956; Myers, 1969; Morgan, 1970; Shields, 1971), although their citation here must not be taken as an endorsement. The location for earthworm containers depends upon the climate of the region. Outdoor containers or pit-runs (benches) in northern areas will require insulation during the winter period when the soil is normally frozen. Smaller wooden pit-runs, or one of the various types of metal tubs, can be housed in a basement or shed to avoid winter freezing problems. Since the indoor facilities permit year-round activity, these can be a source of replenishment for outside gardens, compost piles, flower beds or earthworm beds, etc. The size of the container can vary. A convenient size is a box 50 cm long × 35 cm wide and 15–20 cm deep. Larger containers, when filled with medium and earthworms, will be extremely hard to move. These boxes should have holes 0.5 cm in diameter drilled in the bottom. Plastic window screening should be placed on the inside bottom of the box with a burlap lining on top of the screen and sides of the box before the soil is added. This permits the excess water to drain and prevents the soil medium from sticking to the box, and also prevents the earthworms from escaping through the holes.

Various combinations of soil and organic matter can serve as a medium in which to raise earthworms. A frequently used mixture is ⅓ soil and ⅔ organic matter. Sources of suitable organic matter are: decayed sawdust, hay, leaves, manure, peat moss, sod, or straw. Additional materials which can be added to the medium to serve as food sources are: chicken starter, cornmeal, and kitchen scraps and fats. Earthworms are omnivorous and can utilize many materials as food sources. Some important facts to remember are: 1) the medium should contain sufficient organic matter so that it will not pack into a dense, soggy mass, 2) the containers must not be overwatered, and 3) the presence of low-watt bright white or blue light will prevent the earthworms from crawling on the surface of the medium and eventually out of the box.

The species most frequently used as fish bait, and therefore the ones most likely to be cultured, are: *Aporrectodea trapezoides, Ap. tuberculata, Ap. turgida,* and *Eisenia foetida.* Two other species, *Lumbricus rubellus* and *Octolasion tyrtaeum,* have also been sold or reared as fish bait, though not so commonly as the others mentioned. The night-crawler *Lumbricus terrestris* is widely used by fishermen but cannot be commercially cultured economically because of its long life cycle, low reproductive rate, and large spatial requirements.

Methods of Study

Sampling Techniques

There are numerous methods for sampling earthworm populations. These fall mainly under the general categories of hand sorting, chemical extraction, electrical extraction, and vibration methods. The effectiveness of these methods depends upon the species and habitat; no one method is equally suitable for all species and all habitats.

Digging and hand sorting is the most reliable sampling method, and the one used primarily to obtain the specimens for this study (Low, 1955; Reynoldson, 1955; Satchell, 1955, 1967, 1969; Svendsen, 1955; Nelson and Satchell, 1962; Zicsi, 1962). Though laborious, digging and hand sorting have been widely used for sampling earthworms and for assessing the effectiveness of other methods. Digging to locate earthworms should be done with two factors in mind, moisture and organic matter (cf. Reynolds and Jordan, 1975), and collecting success will be high if one concentrates on sites where both are present. The digging can be done with a variety of tools—shovel, trowel, garden fork, soil cores, etc. The soil can then be pressed and passed through the fingers, or sieves may be employed. The advantages of this method are two-fold: within a sample area active individuals, aestivating individuals, and cocoons may be collected, and, in addition, a well-defined sampling area may be chosen so that quantitative data may be obtained. There are some disadvantages, however. The method is laborious and time-consuming, specimens less than 2 cm in length may escape collection, and, if digging is restricted to the top layers of soil, very large individuals may escape into the deeper layers. Furthermore, specimens may be damaged and there is considerable habitat destruction.

Chemical extraction is a method widely used to collect earthworms and was a second method employed in the present study. Initial studies on chemical extraction were done by Evans and Guild (1947) using potassium permanganate solution to expel earthworms from the soil. Further experiments with chemical extraction, notably using formalin, were conducted by Raw (1959) and Waters (1955). The standardized sampling format that I have employed over the years of quantitative extraction is based on a $0.25m^2$ quadrat of soil surface. A solution of 25 ml of formalin (37% Formaldehyde Solution, U.S.P.) in four and a half litres of water is sprinkled over each quadrat so that all of it infiltrates the soil without runoff. The earthworms that surface in the ten minutes following the application of the expellant are collected. If the collection is to be obtained for other than scientific purposes (e.g., for bait), the time, strength, and number of applications can be varied, but it should be noted that solutions stronger than 15 ml formalin per litre of water may kill the grass in lawns, and if specimens are to be kept alive for more than a few minutes they must be washed in fresh water immediately upon surfacing because formalin can act as a vermicide. Other materials such as Mowrah meal have been used to expel earthworms from the soil (Jefferson, 1955). With a chemical extraction method the sampling time and labour are reduced, a well-defined sampling area may be chosen, and there is minimum disturbance of the habitat. The disadvantages of the method are that only active individuals are collected, and not cocoons and aestivating or hibernating individuals, only shallow dwelling species or species with burrow systems are collected, there may be poor penetration of the vermicide when certain soil conditions prevail, and there is a variability of response to the vermicides by different species. The technique is generally good for Lumbricidae but poor for the other families.

Electrical extraction, a method described by several authors (Walton, 1933; Johnstone-Wallace, 1937; Doeksen, 1950; Satchell, 1961), has long been used by fishermen to obtain bait. The method requires a generator and one to three electrodes. The current conducted through the soil acts as an expellant. The advantage of this method is minimal disturbance to the habitat. The disadvantages are the excessive time required per sample, the difficulty of defining the exact limits of the volume of soil treated, and the variability of the physical and chemical properties of the soil (for example, when soil is moist, deep dwelling species will surface, but if the surface soil is dry the earthworms may go deeper into the soil). The use of too much current kills the earthworms near the electrodes, and the response to electricity varies in different species.

Vibration methods, or *mechanical extraction*, are currently limited to the southeastern United States. Various modifications of this technique ("grunting" in Florida and Georgia, and "fiddling" in Arkansas) are employed by fish bait collectors and yield earthworms in amazing quantities (Vail, 1972; Reynolds, 1972d, 1973d). Mechanical stimulation by vibrations seems to have very little effect on the Lumbricidae but it is extremely successful for some Acanthodrilidae and some Megascolecidae. These two latter families are not found either in Canada or in Europe, which may account for omission of this technique in European discussions of earthworm sampling, except for one small note (Edwards, R., 1967). The advantages of mechanical extraction are the minimal habitat destruction and the reduced sampling time required for each sample. The disadvantages are the difficulty of defining the exact volume of soil treated, the effects of the variability of the physical and chemical properties of the soil, and the variable response of the different species.

There are several other sampling methods that may be used. Wet sieving involves washing soil with a jet of water through a series of sieves after the soil samples have been removed from the field (Morris, 1922; Bouché and Beugnot, 1972a). There are no available data on the efficiency of this method. The disadvantages, according to Ladell (1936), are the excessive time required per sample, the inordinate amount of labour in residue separation, and the damage to the specimens during separation.

The flotation method employed by Raw (1960) for extracting the microfauna from soils unsuitable for hand sorting was patterned after a technique designed by Salt and Hollick (1944). Its advantage is that it can be adapted to extract earthworm cocoons; thus, all stages of the population can be sampled.

The heat extraction method operates on the principle of the Baermann funnel (Baermann, 1917) and has been used for extracting small surface dwelling species that are difficult to hand sort. Similar designs employing Tullgren funnels and incandescent lights have been used by this author. Considerable time per sample is required for this method as the soil samples have to be brought in from the field and placed on the wire sieves in the funnels for several hours. The method has limited use for earthworm sampling.

Trapping techniques are unlikely to yield accurate population estimates but do form a useful method of studying the activity patterns where population densities are low (Svendsen, 1957). A mechanized soil washing method, involving rotating containers and standing sieves, was described by Edwards et al. (1970). This method is faster than previous washing techniques and is apparently suitable for most soils.

Several authors have compared and discussed the relative efficiency of extracting earthworms from soil by two or more of the previous methods (Svendsen, 1955; Raw, 1960; Bouché, 1969a; Satchell, 1967, 1969). From my own observations, the choice of chemical, electrical, or mechanical methods for extraction of earthworms from the soil is greatly dependent on the genus and species of earthworm to be collected.

Preservation Techniques

The proper preservation of specimens for identification, shipping, and storage has long been a problem. Few good accounts of preserving techniques are readily available to those who wish to send material to a specialist for examination. One of the best media for earthworm preservation is 10–15% formalin because it hardens the specimens to facilitate handling. Weak alcohol solutions leave the specimens soft and limp while strong alcohol solutions produce an undesirable brittleness. In both cases, alcohol also causes a condition known as "alcohol browning". This condition makes the reporting of colour of preserved alcohol specimens valueless. Generally, formalin does not distort the colour greatly.

A simple and effective technique is to kill the worms by immersing them in 70% ethyl alcohol. When movement stops they are placed on absorbent paper in a straight position, and allowed to dry for a few minutes. For preservation they should then be transferred to a container of 10–15% formalin where they will harden in the position thus placed. They must be straight because curled or twisted specimens are more difficult to handle when internal examination and dissection are required. The specimens should be left in this container overnight and may then be stored in bottles or vials filled with fresh formalin preservative without much danger of curling. For best results, the preservative should be changed again in a week, especially for such species as *Aporrectodea trapezoides, Lumbricus rubellus*, and *L. terrestris*. Diffusion of body fluids from these species to allow replacement with the preservative seems to take a longer period. As a general rule, the preservative should be changed at weekly intervals until it remains clear.

Ontario Collection Coding

Under the Ontario Distribution for each species (Systematic Section), the collection data have been coded in a consistent manner: location, habitat (when available), date of collection, collector(s), number of specimens (by age classification, explained in the Glossary), and museum number (if any). When an author and date are given, the collection information can be found in that source. If the data are given for a literature source, it is because the author has examined that collection. An asterisk (*) before a location means that the author has a 35 mm colour slide of the habitat in his photographic collection. The abbreviations used in the Ontario Distribution records are:

ANSP	Academy of Natural Sciences, Philadelphia
AT	Alexis Troicki
BP	bypass
CNM	National Museums of Canada
CO	county
CWR	Charles W. Reynolds
DIST	district
DPS	Donald P. Schwert
DRB	David R. Barton
DWR	David W. Reynolds
e	east
e.e.	east edge
GE	Gustav Eisen
GM	G. Mueller
GMN	G. Morley Neale
GWA	George W. Abbott
Hwy	highway
IMS	Ian M. Smith
Jct	Junction
JEM	John E. Moore III
JO	Jack Oughton
JPM	J. Percy Moore
JRD	John Richard Dymond
JWR	John W. Reynolds
km	kilometre(s)
LWR	L. Whitney Reynolds
m	metre(s)
mm	millimetre(s)
MKG	Matthew K. Graham
n	north
n.e.	north edge
RC	R. Cain
Rd	road
RGR	Ruth G. Reynolds
RLH	R. Landis Hare
RNS	R.N. Smythe
ROM	Royal Ontario Museum
RVW	R.V. Whelan
s	south
s.e.	south edge
ST	Steve Tilton
St	street
TTRS	Tall Timbers Research Station
TW	Thomas Weir
TY	Toshio Yamamoto
w	west
w.e.	west edge
WMR	Wilma M. Reynolds
USNM	United States National Museum (Smithsonian)
UW	University of Waterloo

Figure Coding

The figures for each species were drawn with a camera lucida from preserved specimens in the author's collection. The source of the specimens for each drawing is given in parentheses after each figure caption. The abbreviations used in all figures are:

a	anus
bc	buccal cavity
cag	calciferous gland
cg	cerebral ganglion
chl	chloragogen cells
cl	clitellum
clm	coelom
cm	circular muscle
cpc	circumpharyngeal connectives
cr	crop
cut	cuticle
dp	dorsal pore
dv	dorsal vessel
epi	epidermis
es	oesophagus
fp	female pore
g	gizzard
gl	gut lumen
GM	genital markings
GS	genital setae
GT	genital tumescence
h	heart
if	intersegmental furrow
int	intestine
lm	longitudinal muscle
lnv	lateral neural vessel
m	mouth
mf	male funnel

mp male pore
mL mid-lateral line
n nephridium
nb nephridial bladder
np nephropore
ns nephrostome
nt nephridial tube
o ovary
od oviduct
os ovisac
ph pharynx
phm pharyngeal muscle
pp periproct
pr prostomium
ps peristomium
ptm peritoneum
s seta
sep septum
sg seminal groove
sm setal muscle
snv subneural vessel
sp spermathecae
spp spermathecal pore
sv seminal vesicle
t testes
TP tubercula pubertatis
typ typhlosole
vd vas deferens
ve vas efferens
vnc ventral nerve cord
vv ventral vessel

General Morphology

The Oligochaeta are defined as annelids with internal and external metameric segmentation throughout the body, without parapodia but possessing setae on all segments except the peristomium and periproct, with a true coelom and closed vascular system, generally hermaphroditic with gonads few in number in specific locations, with special ducts for discharge of genital products, with a clitellum that secretes cocoons in which ova and spermatozoa are deposited, and which are fertilized and develop without a free larval stage.

The following brief discussion refers primarily to the Lumbricidae, which make up nearly all of the Canadian megadrile fauna. The terms used in this section that are not explained in detail in the text will be found in the illustrated Glossary. For additional information and details of megadrile morphology consult Stephenson (1930) or Edwards and Lofty (1972) for English accounts, and Avel (1959) or Bouché (1972) for French.

External Structure

Terrestrial oligochaetes vary greatly in size. *Bimastos* spp. are less than 20 mm long, the largest tropical species are over 1200 mm (*Glossoscolex, Megascolides*), and some Australian forms may reach 3000 mm in length. The largest species in Canada is *Lumbricus terrestris* (p. 99), which varies from 90 to 300 mm when mature. The body shape is generally cylindrical though usually flattened dorsoventrally in the posterior region in the case of burrowing species.

The entire body is divided along the longitudinal axis into segments separated by intersegmental furrows and septa. This is primary segmentation. There are also secondary annuli, or furrows, which appear to subdivide some of the individual segments, usually in the anterior region. These demarcations are only external. Ljungström and Reinecke (1969) have suggested using α and β for these subdivisions and I use γ for a third subdivision; the primary segments are numbered by roman numerals. There is a loss of uniformity in segmentation at the anterior end of the earthworm; this condition is referred to as cephalization (cf. Gates, 1972c). The first body segment, containing the mouth, is known as the peristomium and may have a tongue-like lobe projecting anteriorly. The prostomium is located above the mouth, and is not a true segment. Its appearance is often important in species identification. The last, or caudal, body segment is referred to as the periproct.

Sometimes a swelling may be seen around the body, the clitellum. The layman frequently mistakes this for the scar of a regenerated animal. In fact it is an epidermal modification of sexually mature specimens where gland cells secrete material to form the cocoon.

Characteristic of all earthworms are the short bristles or setae, retractile structures that add to the worm's grip during tunneling and locomotion. The setae are produced by cells in the body wall. In the Lumbricidae and Sparganophilidae there are four pairs of setae per segment, except for the peristomium and periproct, which are asetal. The type and position of these setae have been used as taxonomic characters (see Glossary—setae, setal formula, setal pairings).

The colour of the megadriles is primarily a result of pigment in the body wall.

But it may be a secondary result of lack of pigment and the red colour of some forms is due to haemoglobin in the blood. Some colour is due to the presence of yellow coelomic corpuscles near the surface, but the presence of chloragogen cells near the surface is rarely, if ever, an influence on colour. Preliminary results of current North American studies indicate that the physical and chemical properties of the soil are a possible influence on earthworm colour.

The body wall, upon which the excretory, genital, and reproductive apertures all open, comprises six layers. From the outside these are: cuticle, epidermis, nerve plexus, circular muscle, longitudinal muscle, and peritoneal layer. The well-developed muscle layers are important in locomotion. The body wall is the foundation for many glandular swellings such as the clitellum, tubercula pubertatis, and genital tumescences, all of which have long been employed as taxonomic characters.

Internal Structure

The annelids have often been characterized as possessing a "tube-within-a-tube" body style (Figs. 2 and 3). The outer tube is formed by the body wall and the inner tube by the alimentary canal. Between these two tubes is the secondary body cavity, or coelom, which is divided at each segment by a septum at the intersegmental furrow. Non-segmental alignment may occur anteriorly in some species as a result of cephalization. The coelomic cavity is filled with a fluid that varies in composition interspecifically, and also intraspecifically for those species that are euryecious in that they tolerate a wide range of habitat conditions. Pores in the septa permit the coelomic fluid to pass freely between segments.

The alimentary canal or digestive tract is essentially a tube extending from mouth to anus. The anteriormost part of the tract consists of a muscular buccal cavity, followed by a pharynx which has a sucking action during feeding, the oesophagus, the crop, a crushing organ known as the gizzard, and finally the intestine. The intestine may possess a dorsomedian fold, the typhlosole, that serves to increase the absorptive surface. Many associated structures are connected to the alimentary system, viz., blood glands, chloragogen cells, calciferous glands, and salivary glands. An extensive account of the alimentary canal is found in Gansen (1963).

The circulatory system is closed but there is an extensive sinus between the intestinal epithelium and the choragogen cells. Extending almost the total length of the body are three main vessels (Fig. 3): the dorsal vessel, closely associated with the alimentary canal for most of its length, and two ventral vessels (ventral and sub-neural vessels). The ventral vessel is located between the nerve cord and the alimentary canal, while the sub-neural vessel is located between the nerve cord and the body wall. These main vessels are connected in each segment by paired connectives. In several anterior segments these connectives, termed "hearts", are enlarged and contractile, and possess valves. There are other trunks and branches which anastomose throughout the body. The circulatory or vascular system has not yet achieved its proper position in oligochaete systematics. Its importance has been discussed by Gates (1972c) and Reynolds (1973e).

There is no formalized respiratory system in earthworms; exchange of oxygen and carbon dioxide takes place through the moist cuticle. Respiration normally

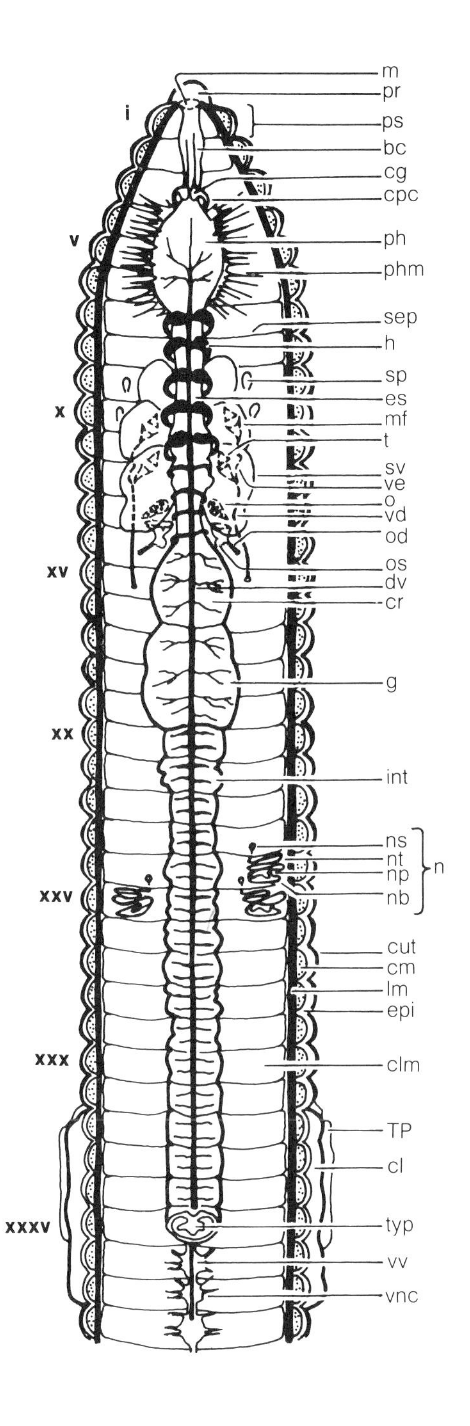

Fig. 2 Diagrammatic longitudinal section of a lumbricid earthworm showing internal organs.

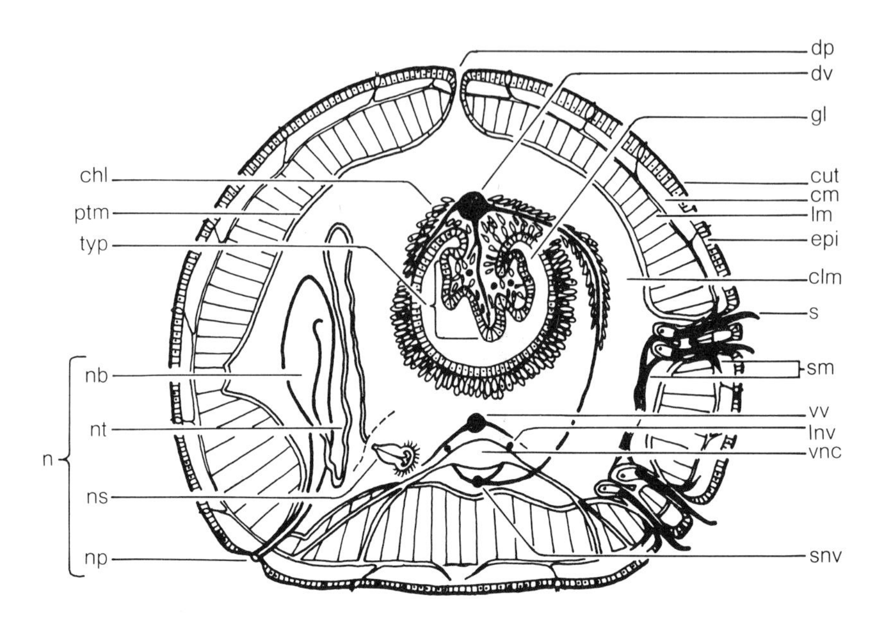

Fig. 3 Diagrammatic cross section of a lumbricid earthworm.

occurs in air but earthworms can exist in water for long periods of time (e.g., six months) if the water is highly oxygenated (Brown, 1944, Roots, 1956).

The excretory system is composed of a series of coiled tubes called nephridia (sing. nephridium). These are the main organs for nitrogenous excretion in earthworms. In the Lumbricidae, they are paired organs in each segment except the first three and the last. A nephridium occupies part of two successive segments where the nephrostome, or funnel, is in the anterior segment and the coiled tube and nephridial bladder are in the posterior segment. The nephridial bladder passes through the body wall opening to the outside forming the nephropore. The position of the nephropore, as well as the structure and type of nephridium, are used as taxonomic characters. The most complete discussion of nephridia and their classification was presented by Bahl (1947).

The nervous system is concentrated, with a bilobed mass of nervous tissue (cerebral ganglia) on the dorsal surface of the pharynx which is connected to subpharyngeal ganglia by a pair of circumpharyngeal connectives. The nerve cord, a fusion of the circumpharyngeal connectives, extends caudad from the subpharyngeal ganglia ventrally between the alimentary canal and the body wall (Fig. 2). In each segment, posterior to iv, a ganglion is formed and three pairs of nerves (peripheral nervous system), one pair anterior to the ganglion and two pairs posterior to the ganglion, extend to the motor and sensory areas. The nervous system is another portion of somatic anatomy that has not yet achieved its proper position in oligochaete systematics.

The reproductive system has long been used as the main source of taxonomic characters. In amphimictic species the male gonads are paired testes found in segments x and xi close to the anterior septa, a condition termed holandric. Anterior to each testis, in segments ix and x, and also posteriorly in segments xi and xii, lobed seminal vesicles occur in which the sperm develop. Sperm are transferred via the sperm funnel and sperm ducts to a vas deferens that may traverse several segments before opening to a male gonopore. The female gonads are represented by a pair of gonads found in segment xiii. Ripe oocytes pass through the coelomic fluid into ovisacs which lead via an oviduct to the female genital pore. In each of segments ix and x there is a pair of sac-like organs, opening ventrally, that receive the sperm during copulation. These are the spermathecae.

In recent years the inadequacies and inconsistencies of reproductive characters in taxonomy have been discussed (Reynolds et al., 1974; Reynolds, 1974c; Gates, 1974a). Unfortunately, statements such as "Oligochaetes are hermaphrodite, and have more complicated genital systems than unisexual animals" (Edwards and Lofty, 1972) are true only in the broadest sense (cf. Reynolds, 1974c). In this study, eight of the 18 lumbricids are parthenogenetic (or unisexual). In megadriles, only the clitellum, ovaries, oviducts, and possibly ovisacs are essential to reproduction (cf. Gates, 1974a; Reynolds, 1974c). Therefore, when reproduction is parthenogenetic all of the following are no longer required: testes, seminal vesicles, seminal receptacles, vas deferens, copulatory chambers, copulatory penes, prostates and ducts, genital markings, spermathecae, tubercula pubertates, genital and penial setae.

The external position and morphology of the genital apertures, setae and tumescences, clitella, and tubercula pubertates have been widely used in lumbricid identification. If these characters are constant for a given species, they are excellent simple characters that non-specialists can use with reliability.

A, B, C, D These single capital letters refer to the meridians of longitude passing anteroposteriorly along the apertures of the respective setal follicles. Thus, *A* represents a line along the *a*, the most ventrally located setal follicles.

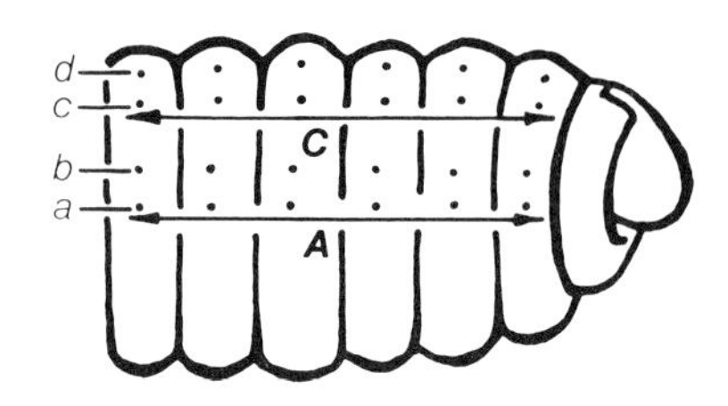

AA, BB, CC, DD See setal formula.

acinus (Fr. *acine* m.) A sac-like termination of a branched gland.

aclitellate adults (Fr. *adultes sans clitellum, antéclitellienne* f.) These are prereproductive individuals without a clitellum but in which genital markings are obvious. The second number in the age classification formula (q.v.) refers to such individuals.

adiverticulate (Fr. *sans diverticule*) Without diverticula, and usually referring to spermathecae.

aestivation (Fr. *estivation* f., *anhydrobiose*) A period of inactivity, or dormancy, resulting from unfavourable moisture conditions.

age classification formula A series of numbers following a binomen (usually three or four numbers) separated by dashes indicating the number of: juveniles—aclitellate adults—clitellate adults—postclitellate adults in a collection. If there are no postclitellate adults in the collection the final zero is omitted from the formula. See juveniles, aclitellate adults, clitellate adults, postclitellate adults.

amphigony See amphimixis.

amphimixis (Fr. *amphimixie* f.) Reproduction involving fertilization of an ovum by a sperm. In megadriles the same as biparental reproduction. Cf. parthenogenesis.

anal segment See periproct.

anastomosis (Fr. *anastomose* f.) Cross connections of ducts, branches of organs, or, more usually, of blood vessels.

anthropochore (Fr. *anthropochore*) Transported by man, usually unintentionally. Cf. peregrine.

aortic arch See hearts.

asetal (Fr. *sans soies*) Without setae. Cf. peristomium, periproct.

atrial gland (Fr. *glande atriole* f.) Glandular tissue associated with a cleft or coelomic invagination containing the male pore.

blood glands (Fr. *glandes sanguines* f.) Follicles clustered in the pharyngeal region, supposed to function in the production of haemoglobin and blood corpuscles.

brain (Fr. *cerveau* m.) See cerebral ganglion.

buccal cavity (Fr. *cavité buccal* f.) (**bc**) The first region of the alimentary canal, between mouth and pharynx (Fig. 2).

C. Abbreviation for circumference (in German publications replaced by U). See setal formula.

caecum (Fr. *caecum* m.) A blind diverticulum or pouch from the alimentary canal.

calciferous gland (Fr. *glande de Morren, glande calciférе* f.) (**cag**) Whitish gland that secretes calcium carbonate and opens into the gut via the oesophageal pouches. In Lumbricidae, it is generally found in segments x–xiv.

castings (Fr. *déjections de surface* f., *turricules* m.) Faeces, the voided earth and other waste matter that are commonly deposited on the surface of the ground. Not all species, however, form their casts above the ground.

cephalization (Fr. *céphalisation* f.) The loss of metameric uniformity at the anterior end of the body.

cerebral ganglion (Fr. *ganglion cérébral* m.) (**cg**) Concentrated nerve cells above the alimentary canal that function as a simple brain (Fig. 2).

cf. (*confer*) Compare.

chaeta See seta.

chloragogen cells (Fr. *cellules chloragogues* f.) (**chl**) Cells surrounding the alimentary canal; their function is uncertain but is attributed to excretion and regeneration in the literature (Fig. 3).

cingulum See clitellum.

circumpharyngeal connective (Fr. *connectif circumpharyngien* m.) (**cpc**) Nerve collar, between cerebral ganglion and ventral nerve ganglion (Fig. 2).

clitellate adult (Fr. *adulte avec clitellum, clitellienne* f.) Those individuals with developed clitellum and genital markings. The third number in the age classification formula (q.v.) refers to these individuals.

clitellum (Fr. *clitellum* m.) **(cl)** A regional epidermal swelling where gland cells secrete material to form the cocoon. There are two types recognizable. An annular clitellum or cingulum (Fr. *anneau* m.) encircles the body whereas a clitellum that encompasses only the dorsal and lateral parts of the body is referred to as a saddle (Fr. *selle* f.). The convention xxvi, xxvii–xxxii, xxxiii means that the clitellum is generally found on segments xxvii–xxxii, but may in some individuals overlap onto segments xxvi and/or xxxiii.

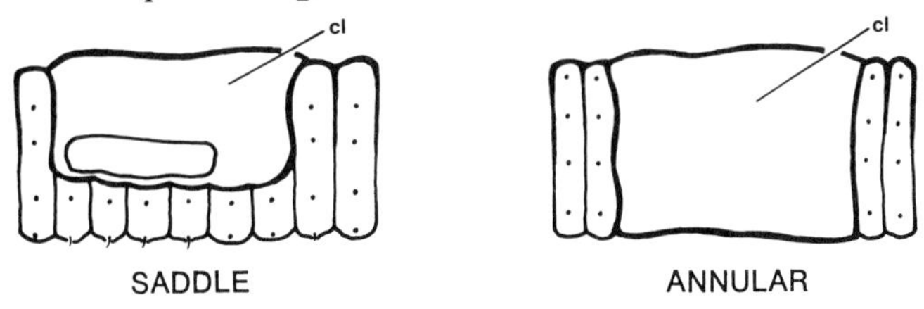

In the case of *Eisenia rosea* the clitellum has been termed flared. This ventral flared condition is easily recognizable.

coelom (Fr. *cavité coelomique, coelome* f.) **(clm)** The body cavity between the body wall and the alimentary canal (Fig. 3).

congeneric (Fr. *congénère*) Belonging to the same genus.

copulation (Fr. *accouplement* m., *copulation* f.) Sexual union, mating.

crop (Fr. *jabot* m.) **(cr)** A widened portion of the digestive system that lacks the muscularity of the gizzard, in Lumbricidae anterior to the gizzard and posterior to the oesophagus (Fig. 2).

cuticle (Fr. *cuticule* f.) **(cut)** A thin, non-cellular, colourless, transparent outer layer of the body wall. See iridescence 2.

diapause (Fr. *diapause* f.) An obligatory resting stage in development.

digitiform (Fr. *digitiforme*) Finger-shaped.

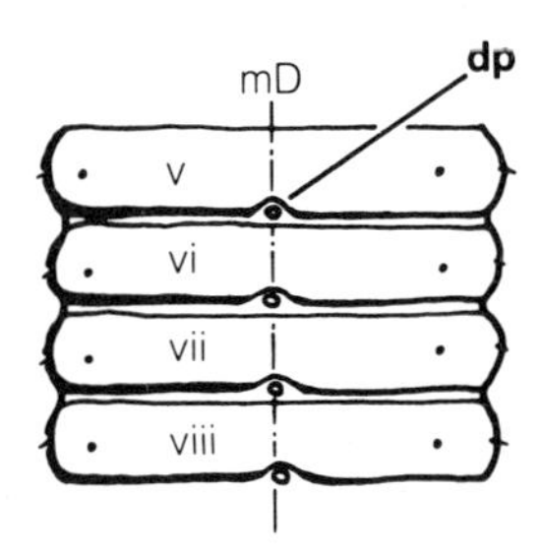

dorsal pore (Fr. *pore dorsal* m.) **(dp)** Small single intersegmental apertures in the mid-dorsal line (mD) leading to the coelomic cavity (Fig. 3). The convention first dorsal pore 5/6 means that the dorsal pore is found in the intersegmental furrow between segments v and vi.

dorsal vessel (Fr. *vaisseau dorsal* m.) **(dv)** A major blood vessel located above the dorsal surface of the alimentary canal (Figs. 2, 3).

ectal Outer, external, toward the body wall.

egg sac See ovisac.

endemic (Fr. *endemique*) Restricted to a certain region or part of a region, native. Cf. exotic, indigenous.

ental Inner, internal, away from the body wall.

epidermis (Fr. *epiderme* m.) **(epi)** The outer cellular layer of the body wall, which secretes a protective cuticle (Fig. 3).

epilobic (Fr. *epilobique*) See prostomium.

eq. Equatorial, see mL.

euryoecious (Fr. *euryoeciques*) Having a wide range of habitat tolerance.

exoic (Fr. *exoique*) Opening to the exterior through the epidermis, referring to the excretory system.

exotic (Fr. *exotique*) Introduced, foreign. Cf. endemic, indigenous.

facultative (Fr. *facultatif*) Conditional, having the power to live under different conditions. Cf. obligatory.

female ducts Gonoducts. See oviducts.

female pores (Fr. *pores femelles* m.) **(fp)** The external openings for the oviducts on segment xiv (Lumbricidae) and ventrad of the mid-lateral line. They are usually more difficult to see than the male pores.

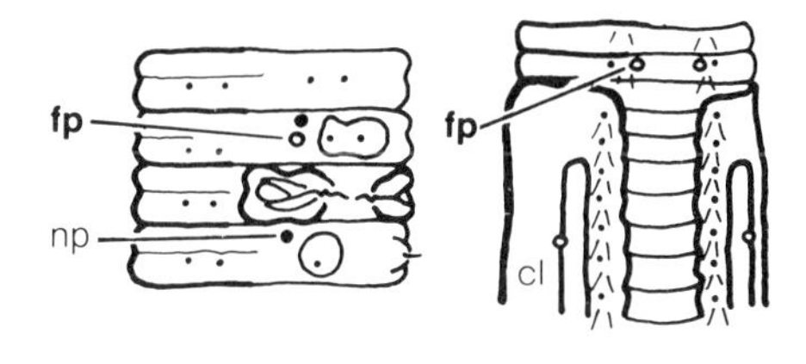

fide On the authority of, or with reference to publication, to a cited published statement.

flared clitellum (Fr. *clitellum évassé* m.) See clitellum.

genital markings (Fr. *mamelons antiarrheniques, mamelons periarrheniques* m.) **(GM)** Glandular swellings, pits or grooves of the epidermis. See genital tumescences.

genital setae (Fr. *soies genitales* f.) **(GS)** See setae.

genital tumescences (Fr. *papille puberculienne* f.) **(GT)** In Lumbricidae, areas of modified epidermis (glandular swellings) without distinct boundaries and through which follicles of genital setae open.

girdle See clitellum.

gizzard (Fr. *gésier* m.) **(g)** The muscularized portion of the digestive system, in Lumbricidae, anterior to the intestine and posterior to the crop (Fig. 2).

gonopore (Fr. *gonopore* m.) See male pores, female pores.

hearts (Fr. *coeurs* m.) **(h)** The enlarged, segmental, pulsating connectives of the blood system between the ventral and one or two other longitudinal trunks (e.g., dorsal and/or supra-oesophageal) (Fig. 2).

hemerobiont A species dependent on human culture.

hemerodiaphore A species indifferent to the influence of human culture.

hemerophile A species favoured by human culture.

hemerophobe A species averse to the influence of human culture.

hibernation (Fr. *hibernation* f.) A period of inactivity or dormancy resulting from unfavourable temperature conditions.

holandric (Fr. *holandrique*) The condition where the testes are restricted to segments x and xi, or a homoeotic equivalent.

holoic (Fr. *holonéphridique*) The condition of having a pair of stomate, exoic nephridia in each segment of the body except the first and last.

homoeotic (Fr. *homoeotique*) The condition of having glands or organs in a segment(s) where they do not normally occur. Refers principally to intraspecific variation.

indigenous (Fr. *indigène*) Belonging to a locality, not imported, native. Cf. endemic, exotic.

in litt. (*in litteris*) In correspondence.

intersegmental furrow (Fr. *sillon intersegmentaire* m.) **(if)** The boundary between two consecutive segments; the area where the epidermis is thinnest and where, in pigmented species, colour is lacking.

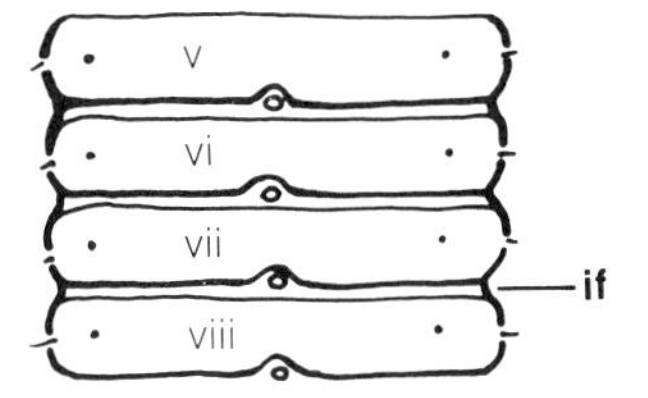

iridescence (Fr. *irisation, iridescence* f.) In the context of earthworm biology this refers to 1) the appearance of sperm aggregated on the male funnels (q.v.), or 2) the appearance of cuticular colour as a result of refracted light.

juveniles (Fr. *larves* f.) Those individuals with no recognizable genital markings such as the clitellum, tubercula pubertatis, tumescences, etc., i.e., in the life stage between hatching and the appearance of genital markings. The first number in the age classification formula (q.v.) refers to these individuals.

lamella (Fr. *lamelle* f.) Any thin plate- or scale-like structure.

mD (Fr. *médio-dorsale*) Mid-dorsal line.

mL (Fr. *médio-latérale*) Mid-lateral line.

mV (Fr. *médio-ventrale*) Mid-ventral line.

male funnel (Fr. *entonnir mâle* m.) **(mf)** The enlargement of the ental end of a sperm duct with a central aperture through which sperm pass into the lumen of the duct on their way to the exterior. Sperm may temporarily aggregate on the funnels, prior to entering the ducts, their presence being indicated by iridescence (q.v.).

male pores (Fr. *pores mâles* m.) **(mp)** The external openings for the male ducts through which sperm are liberated during copulation. In Lumbricidae they are usually conspicuous near the mL on segment xv; any variation is noted in the diagnosis.

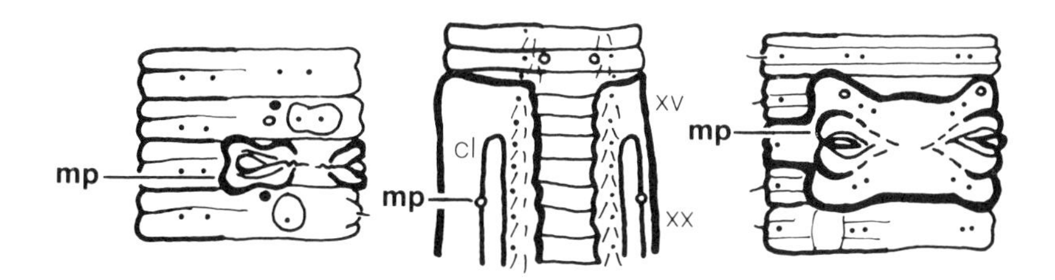

male sterility (Fr. *sterilité mâle* f.) Often cited as evidence for parthenogenesis (q.v.) and may be indicated by the following: 1) adult retention of juvenile testes, 2) adults with juvenile seminal vesicles and no evidence of sperm, 3) the absence at maturity of iridescence on the male funnels, indicating that there are no mature sperm aggregations, 4) the absence of similar iridescences in the male ducts and/or spermathecae, and 5) the absence of externally adhesive spermatophores. These criteria will only suggest male sterility in any given individual and many cases of repeated evidence are required before a species can be considered male sterile or parthenogenetic.

megadrile (Fr. *mégadrile* m.) Sensu Gates (1972c: 29) and Reynolds and Cook (1977), this term is synonymous with terrestrial oligochaetes. There is some morphological basis for the megadrile/microdrile division of the Oligochaeta (cf. Gates, 1972c). Brinkhurst (*in* Brinkhurst and Jamieson, 1971: 104) employs microdrile as a major heading when discussing the aquatic oligochaetes. In general, these old terms are used to describe terrestrial and aquatic oligochaetes without any systematic judgments.

mesial (Fr. *medial*) In the middle vertical or longitudinal plane.

metamere (Fr. *metamere* m.) A segment.

moniliform (Fr. *moniliforme*) Arranged like a string of beads.

monotypy (Fr. *monotypie* f.) The situation arising when a genus-group taxon is established with only one originally included species; or when a family-group taxon is established with only one originally included genus.

morph (Fr. *forme* f., *morph* f.) A group of individuals that share a common anatomy resulting from degradations, deletions, or other changes from structure of the ancestral amphimictic population caused by reproductive isolation. Such isolation usually comes about as a result of parthogenesis.

Morren's gland See calciferous gland.

mouth (Fr. *bouche* f.) **(m)** The anterior opening to the alimentary canal located in the peristomium.

mouth cavity See buccal cavity.

muscular tube See nephridial bladder.

nearctic (Fr. *néarctique*) A zoogeographical region including Canada, the United States, Greenland, and northern Mexico.

neotype (Fr. *neotype* m.) A single specimen designated as the type specimen of a nominal species-group taxon of which the holotype (or lectotype), and all paratypes or all syntypes are lost or destroyed. Neotypification is the act of selecting a neotype. (For nominal taxon, see taxon.)

nephridial bladder (Fr. *vesicule de la nephridie* f.) (**nb**) The extended portion of the nephridial tube connected to the nephropore (Fig. 3).

nephridial pore See nephropore.

nephridial reservoir See nephridial bladder.

nephridiopore See nephropore.

nephridium (pl. **nephridia**) (Fr. *nephridie* f.) (**n**) The organ for nitrogenous excretion (Figs. 2, 3).

nephropore (Fr. *nephridiopore* m.) (**np**) The external opening of a nephridium (Fig. 3).

nephrostome (Fr. *nephrostome* m.) (**ns**) The ciliated funnel at the ental end of the nephridium (Fig. 3).

obligatory (Fr. *obligatoire*) Limited to one mode of life or action. Cf. facultative.

oesophagus (Fr. *oesophage* m.) (**es**) The portion of the gut between the pharynx (anterior) and crop (posterior), ending in an oesophageal valve (Fig. 2).

omnivorous (Fr. *omnivore*) Eating both animal and plant tissue.

op. cit. (*opere citato*) In the work or article previously cited for this writer (no page cited).

ovary (Fr. *ovaire* m.) (**o**) The organ for ova (egg) production (Fig. 2).

oviducal pores See female pores.

oviduct (Fr. *oviducte* m.) (**od**) The duct carrying the ova from the coelomic funnel to the exterior (Fig. 2).

ovisac (Fr. *ovisac* m.) (**os**) An egg-capsule or receptacle (Fig. 2).

ovum (pl. **ova**) (Fr. *ovule, oeuf* m.) The female germ cell, matured egg-cell.

palaearctic (Fr. *paléoarctique*) A zoogeographical region including all of Europe and the U.S.S.R. to the Pacific Ocean, Africa north of the Sahara Desert, and Asia north of the Himalaya Mountains.

papilla (Fr. *papille* f.) A protruding dermal structure.

parietes (Fr. *pariétes* m.) Walls or sides of structures.

parthenogenesis (Fr. *parthénogénèse* f.) Uniparental reproduction in which the ova develop without fertilization by spermatozoa. Cf. amphimictic.

penial setae (Fr. *soies de la verge* m.) See seta.

peregrine (Fr. *peregrin*) Widely distributed, not necessarily involving man.

periproct (Fr. *pygidium* m.) **(pp)** The terminal (last, caudal) "segment" of the body, without coelomic cavity, asetal.

peristomium (Fr. *peristomium* m.) **(ps)** The first body segment, asetal, and containing the mouth (Fig. 2).

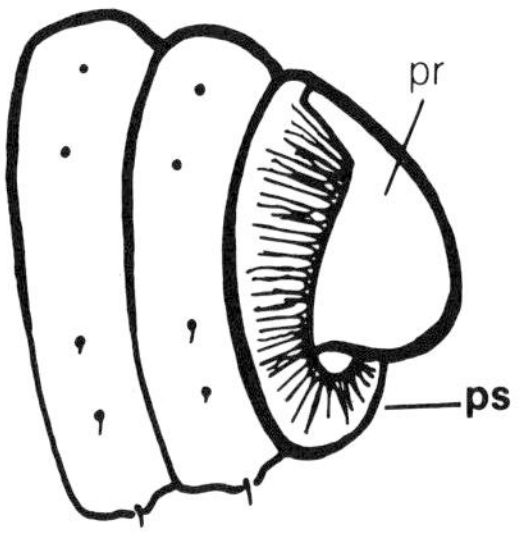

pr
ps

pH (Fr. *pH* m.) An indication of acidity or alkalinity measured as the negative logarithm of the hydrogen-ion concentration, and expressed in terms of the pH scale (0–14) where pH 7 is neutral, less than 7 is acidic, and more than 7 is alkaline. Previously, North American studies employed an aqueous solution to make soil pH readings, and these are the figures given in the text, but variations can occur when the amount of water present in the soil changes as well as when the amount of dissolved gases in this water, e.g., CO_2, changes. To overcome these variations in the pH readings, one of several salt solutions of differing strengths may be employed instead of water, e.g., KCl or $CaCl_2$. (For details, see Peech, 1965.)

pharynx (Fr. *pharynx* m.) **(ph)** The portion of the gut between the buccal cavity (anterior) and the oesophagus (posterior) (Fig. 2).

pinnate (Fr. *penne*) Divided in a feathery manner.

polymorphism (Fr. *polymorphisme* m.) Occurrence of different forms of individuals within the same species.

postclitellate adult (Fr. *adulte après clitellum, postclitellienne* f.) Postreproductive individuals without a clitellum but with areas of discolouration in the regions of the clitellum, and with genital markings. If these discolourations disappear (which is not abnormal), differentiation between aclitellate adults and postclitellate adults may be impossible even after dissection. These individuals have reverted to an aclitellate state and in the future may become clitellate again and be reproductive. The fourth number in the age classification formula refers to these individuals, but if such individuals are not present in the sample then this fourth figure is omitted instead of using a zero.

prostates (Fr. *prostates* f.) In Lumbricidae, the same as atrial glands, and of unknown function.

prostatic pores (Fr. *pores prostatique* m.) See male pores.

prostomium (Fr. *prostomium* m.) **(pr)** The anterior lobe projecting in front of the peristomium and above the mouth. There are four types as seen in dorsal view:

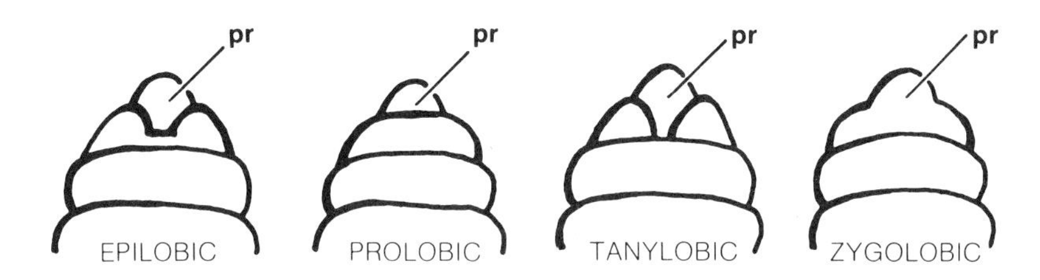

1) Epilobic: tongue of the prostomium partly divides the peristomium. 2) Prolobic: prostomium demarcated from the peristomium without a tongue. 3) Tanylobic: with a tongue that completely divides the peristomium. 4) Zygolobic: prostomium not demarcated in any way.

pseudogamy (Fr. *pseudogamy*) The activitation of ova by a sperm without nuclear fusion and thus without true fertilization.

pygidium See periproct.

pygomere See periproct.

pyriform (Fr. *pyriforme*) Pear-shaped.

quiescence (Fr. *quiescence* f.) A period of inactivity, or dormancy, resulting from an unfavourable environment; cf. aestivation and hibernation.

q.v. (*quod vide*) Which see.

ridge of puberty (Fr. *crêtes de puberté* f.) See tubercula pubertatis.

sacculate (Fr. *saccule* m.) Provided with sacculi, small sacs or pouches.

saddle See clitellum.

secondary annulation (Fr. *sillons transversaux* m.) **(sa)** The furrows which occur between the intersegmental furrows (q.v.). These demarcations are only external and are labelled α, β, or γ.

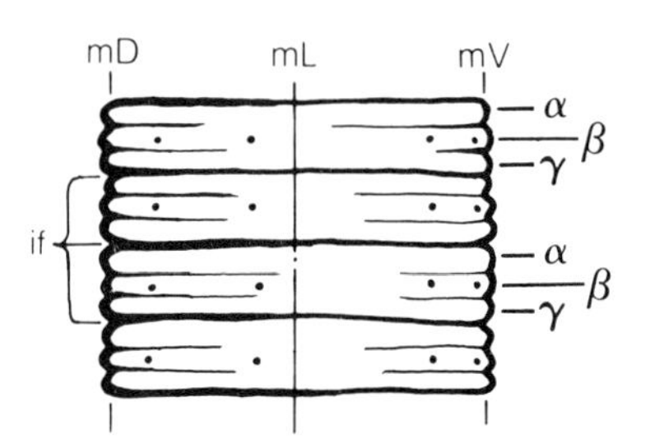

segment (Fr. *segment* m.) A portion of the body, along the anteroposterior axis, between two consecutive intersegmental furrows and the associated septa. Segments are numbered with lower case roman numerals, i, ii, iii, etc., beginning anteriorly with the peristomium as i. The older system and some microdrile workers used upper case numerals, I, II, III, etc.

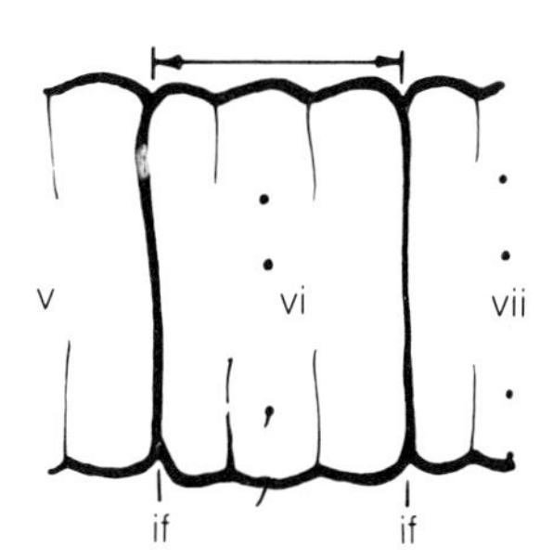

seminal receptacles See spermathecae.

seminal reservoirs See seminal vesicles.

seminal vesicles (Fr. *vésicules séminales* f.) **(sv)** The storage sacs for an earthworm's own sperm until copulation.

septum (pl. **septa**) (Fr. *cloison* f.) **(sep)** The internal partition at intersegmental furrows. Also acts as a supporting membrane for internal organs (Fig. 2).

seta (Fr. *soie* f.) **(s)** A solid rod or bristle secreted by cells at the ental end of a tubular epidermal ingrowth, the setal follicle. Setae are of several types: 1) general: sigmoid shape with pointed outer tip; 2) genital: associated with genital tumescences and/or gonopores, and not sigmoid; 3) penial: associated with the male pores and not sigmoid. Individual setae are referred to as *a, b, c, d*, as shown in the first diagram of this Glossary, *a* being the most ventral and *d* the most lateral of the setae on a particular segment.

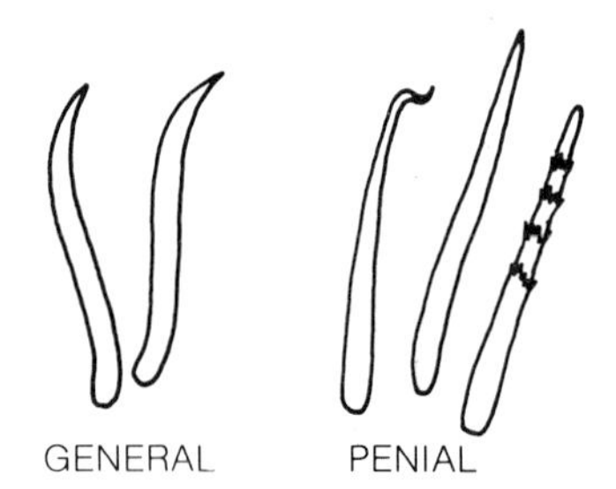

setal formula (Fr. *des soies* f.) The distance between the setae, usually measured on segments x and/or xxx, and being an estimate of the space between the *A, B, C*, and *D* meridians (q.v.). The data can be expressed as a ratio (e.g., $AA{:}AB{:}BC{:}CD{:}DD = 9{:}3{:}6{:}2{:}30$), as groupings (e.g., $AA > BC < DC$, $AA = BC$) or in terms of the circumference, C, (e.g., $DD = 1/2C$). See also setal pairings.

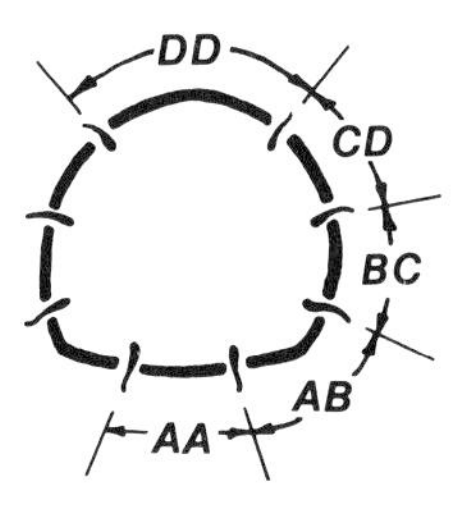

setal pairings (Fr. *schéma de la disposition des soies*) Setae may be closely paired (Fr. *soies etroitement géminées*), widely paired (Fr. *soies distantes*), or separate (Fr. *soies écartées, soies séparées*).

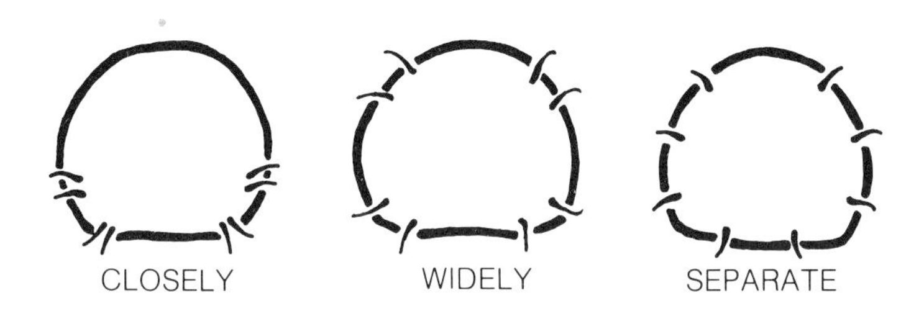

somatic (Fr. *somatique*) Referring to any portion of the anatomy except the reproductive organs.

sperm (Fr. *spermatozoides* m., *sperme* m.) The male germ cells, fertilizing agent.

sperm ducts See vas deferens.

sperm funnel See male funnel.

sperm sacs See seminal vesicles.

spermathecae (Fr. *spermatheque* f.) **(sp)** The pouches developed in the septa which receive sperm from another individual during copulation; the sperm are stored here until the period of cocoon laying.

spermatophore (Fr. *spermatophore* m.) A capsule of albuminous matter containing a number of sperm.

spermatozoa See sperm.

spermiducal pores See male pores.

stomate (Fr. *stomate*) Referring to open nephridia, i.e., with funnel.

tanylobic (Fr. *tanylobique*) See prostomium.

taxon (pl. **taxa**) (Fr. *taxon* m.) Any taxonomic unit such as a particular family, genus, or species. **Nominal taxon:** The taxon, as objectively defined by its type, to which any given name whether valid or invalid applies.

testis (pl. **testes**) (Fr. *testicules* m.) (**t**) The organs for sperm production.

testis sac (Fr. *sac du testicule* m.) Usually a closed off coelomic space containing one or both testes and male funnels of a segment.

trabeculate (Fr. *trabeculaire*) Seminal vesicles that develop as connective tissue proliferations from a septum so as to have numerous irregular spaces that remain inconsiderable until spermatogonia (primitive sperm cells) begin to enter.

tubercula pubertatis (Fr. *puberculum* m.) (**TP**) A glandular swelling appearing near the ventrolateral margins of the clitellum. It is not always present, and it may be continuous or discontinuous, and of varied size and shape.

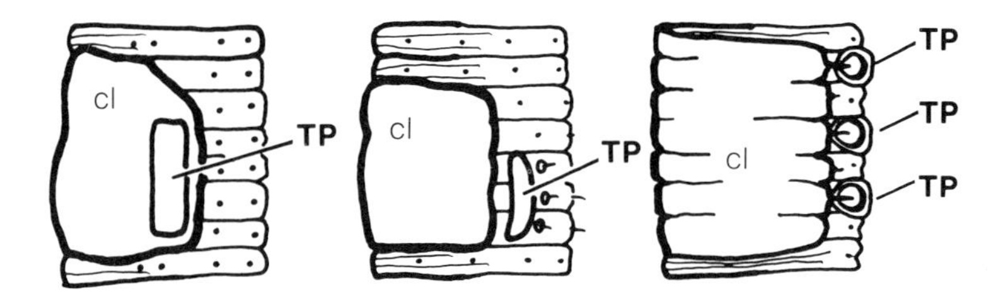

typholosole (Fr. *typhlosolis* m.) (**typ**) Any longitudinal fold in the gut wall projecting into the gut lumen, usually at mD or mV (Figs. 2, 3).

vas deferens (Fr. *canal deferent* m.) (**vd**) The ducts that carry sperm from the male funnels to the exterior (Fig. 2).

ventral vessel (Fr. *vaisseau ventral* m.) (**vv**) A major blood vessel, located ventral to the alimentary canal and dorsal to the ventral nerve cord (Fig. 2).

vesiculate (Fr. *vesiculeux*) Having a vesicle or small bladder-like sac.

viz. (*videlicet*) Namely.

zygolobic See prostomium.

1-1-1-1 See age classification formula.

1/2 See first dorsal pore.

i, ii, iii See segment.

Identification of the Earthworms of Ontario

The identification of the local earthworms is not as difficult as most people suspect. The following key, which should be used in conjunction with the Glossary and diagnoses, has been designed to facilitate identification of the 19 species recorded in Canada without the necessity of dissecting the specimens. Generally speaking, mature specimens are essential for a definitive identification by the non-specialist.

The most useful characters of the key are the nature of the prostomium (zygolobic only in *Sparganophilus eiseni*, tanylobic in species of the genus *Lumbricus*, and epilobic in all others), the segmental position of the clitellum and tuberculata pubertatis (remembering that the prostomium is not numbered), the arrangements of the setae, and the presence or absence of pigment. These characters are all readily visible in fresh material.

If only preserved material is available, some difficulty may be experienced in those parts of the key (couplets 8 and 12) that rely on assessments of colour. In these circumstances it may be necessary at the appropriate point to refer to the detailed diagnoses of several species before proceeding further. At couplet 8, for example, if the colour cannot be assessed confidently it will be necessary to refer separately to the diagnoses of *Dendrobaena octaedra, Dendrodrilus rubidus*, and *Eisenia foetida*. If the preserved specimen at hand definitely is not one of these three species then one continues through the key on the assumption that there is no red pigment present.

The study of the characters used in the key requires no more than a good hand lens or a low-power binocular microscope. The key itself is strictly dichotomous. The numbers in parentheses after the main couplet numbers indicate the couplet from which that particular point in the key was reached. They are inserted to make it easier to retrace one's steps through the key in the event that an obviously incorrect alternative has been reached.

Key to Sexually Mature Earthworms Found in Ontario

1 Prostomium zygolobic; clitellum begins in front of segment xx .... *Sparganophilus eiseni* (p.113)
Prostomium not zygolobic; clitellum begins behind segment xx 2

2 (1) Prostomium tanylobic 3
Prostomium epilobic 6

3 (2) Clitellum begins in front of segment xxx 4
Clitellum begins behind segment xxx 5

4 (3) Clitellum on segments xxvi, xxvii–xxxi, xxxii; tuberculata pubertatis on segments xxviii–xxxi .... *Lumbricus rubellus* (p. 94)
Clitellum on segments xxvii–xxxiii; tuberculata pubertatis on segments xxix–xxxii .... *Lumbricus castaneus* (p. 89)

5 (3) Clitellum on segments xxxi, xxxii–xxxvii; tuberculata pubertatis on segments xxxiii–xxxvi .... *Lumbricus terrestris* (p. 99)
Clitellum on segments xxxiv–xxxix; tuberculata pubertatis on segments xxxv–xxxvii, xxxviii .... *Lumbricus festivus* (p. 92)

6 (2) Clitellum on segments xxiv–xxxi, tuberculata pubertatis absent ...... *Bimastos parvus* (p. 61)
Clitellum rarely on xxiv–xxxi only; tuberculata pubertatis present 7

7 (6) Tuberculata pubertatis small sucker-like discs, usually on segments xxxi, xxxiii, and xxxv (Fig. 4); clitellum on xxviii, xxix–xxxvii .... *Allolobophora chlorotica* (p. 36)
Tuberculata pubertatis not as above 8

8 (7) Pigment present, red 9
Pigment absent, or if present not red 11

9 (8) Setae closely paired; clitellum on segments xxiv, xxv, xxvi–xxxii; tuberculata pubertatis on segments xxviii–xxx; sometimes striped (alternate transverse dark and light bands) .... *Eisenia foetida* (p. 74)
Setae widely paired or separate 10

10 (9) Clitellum on segments xxvii, xxviii–xxxiii, xxxiv; tuberculata pubertatis on segments xxxi–xxxiii, usually darkly pigmented .... *Dendrobaena octaedra* (p. 65)
Clitellum on segments xxvi, xxvii–xxxi, xxxii; tuberculata pubertatis on segments xxviii, xxix–xxx .... *Dendrodrilus rubidus** (p. 69)

11 (8) Male pores equatorial on xiii; female pores on xiv; often yellowish in colour .... *Eiseniella tetraedra* (part) (p. 84)
Male pores not on xiii 12

12 (11) Pigmented 13
Not pigmented 15

13 (12) Setae widely paired; clitellum on segments xxii, xxiii–xxvi, xxvii; tuberculata pubertatis on segments xxiii, xxiv–xxv, xxvi .... *Eiseniella tetraedra* (part) (p. 84)
Setae closely paired 14

14 (13) Clitellum on segments xxvi, xxvii, xxviii–xxxiv; tuberculata pubertatis on segments xxxi–xxxiii .... *Aporrectodea trapezoides* (part) (p. 46)
Clitellum on segments xxvii, xxviii–xxxiv, xxxv; tuberculata pubertatis on segments xxxii–xxxiv .... *Aporrectodea longa* (p. 43)

---

* A rare morph of this species, found in the southern USA, lacks tuberculata pubertatis.

15 (12) Setae widely paired (posteriorly at least) 16
Setae closely paired 17

16 (15) Clitellum on segments xxix–xxxiv; tuberculata pubertatis on segments xxx–xxxiii .... *Octolasion cyaneum* (p. 105)
Clitellum on segments xxx–xxxv; tuberculata pubertatis on segments xxxi–xxxiv .... *Octolasion tyrtaeum* (p. 108)

17 (15) Clitellar region often flared ventrally (bell-shaped in cross section, see glossary, p. 20); clitellum on segments xxv, xxvi–xxxiii; tuberculata pubertatis on segments xxix-xxxi .... *Eisenia rosea* (p. 78)
Clitellar region not flared ventrally 18

18(17) Clitellum begins in front of segments xxx 19
Clitellum begins behind segment xxx; clitellum on segments xxxiii, xxxiv–xlii, xliii .... *Aporrectodea icterica* (p. 40)

19 (18) Genital tumescences often present on segment xxviii, present or absent on segments xxxiii–xxxiv; clitellum on segments xxvii, xxviii–xxxiv; tuberculata pubertatis on segments xxxi–xxxiii; male sterile (see glossary, p. 24) .... *Aporrectodea trapezoides* (part) (p. 46)
Genital tumescences not present on segment xxviii; male fertile 20

20 (19) Clitellum on segments xxvii–xxxiv; tuberculata pubertatis on segments xxxi–xxxiii; genital tumescences often present on segments xxvi and usually absent on segment xxxiii .... *Aporrectodea tuberculata* (p. 50)
Clitellum on segments xxvii, xxviii, xxix–xxxiv, xxxv, tuberculata pubertatis on segments xxxi–xxxiii; genital tumescences often present on segments xxvii and xxxiii .... *Aporrectodea turgida* (p. 56)

Systematic Section

There are three levels of taxa reported in this section—families, genera, and species.

For each family (Lumbricidae and Sparganophilidae), a diagnosis and the designation of the type genus are given. For each of the ten genera occurring in Ontario there are presented a synonymy, the type species, a diagnosis, and a discussion.

The information for the 19 species reported from Ontario is presented in the following order: common name (English and French), synonymy, diagnosis, external anterior illustrations (lateral and ventral views), discussion, biology (habitats, reproduction, etc.), range, North American distribution, Ontario distribution records, and map. For some species, at the end of the North American distribution records are one or more regions listed under "new records". This means that the author has recently received or collected specimens from these areas. Since there are no published records of the species from these areas, they are included without citation to give the reader the most complete information on the distribution of the species in North America. The abbreviations used for the Ontario distribution records are found in the section Methods of Study (p. 10). Niagara County is the union of the former Lincoln and Welland counties. For the purposes of this study North American distribution includes Canada, the United States of America, and Greenland.

It should be noted that unless otherwise stated all biological notes for the species are compilations from the literature. Very little work has been done on the biology of earthworms in Canada and most data are from Europe. The most thorough reviews to which reference may be made for further information are those of Bouché (1972) and Gates (1972c).

Family LUMBRICIDAE Claus, 1880*

Diagnosis

Digestive system: with an intramural calciferous gland comprising longitudinal chambers that open at their anterior ends into the oesophageal lumen, a terminal oesophageal valve reaching into xv, an intestine beginning with a "crop" followed by a gizzard, a sacculated as well as an unsacculated portion and ending in an atyphlosolate region, but without intestinal caeca and supra-intestinal glands. Vascular system: with complete dorsal, ventral, and subneural (and lateroneural?) trunks, the latter adherent to nerve cord, extraoesophageal trunks median to the hearts passing to dorsal trunk in region of x–xii, without supra-oesophageal and lateroparietal trunks. Hearts: lateral, the last pair anterior to segment xii. Nephridia: holoic, vesiculate, ducts passing into parietes in

*Designated by Sims (1973).

region of *B*. Setae, sigmoid and single pointed, eight per segment, in regular longitudinal ranks, in genital tumescences elongated but slender and longitudinally grooved ectally. Dorsal pores, present. Prostomium epilobic, prolobic, or tanylobic. Reproductive system: apertures, all minute, female pores anterior to the male pores, equatorial and anterior to the multilayered clitellum which is always behind xvii. Spermathecae, adiverticulate, pores at intersegmental levels. Ovaries, in xiii, bandlike, each terminating distally in a single eggstring. Ovisacs, in xiv, small, lobed. Ova, not yolky. Prostates, none (after Gates, 1972c: 61–62).

Type Genus
Lumbricus Linnaeus, 1758 (neotypification Sims (1973)).

Genus *Allolobophora* Eisen, 1873

1873 *Allolobophora* Eisen, Öfv. Vet.-Akad. Förh. Stockholm 30(8): 46.
1900 *Allolobophora*–Michaelsen, Das Tierreich, Oligochaeta 10: 480.
1910 *Allolobophora*–Michaelsen, Ann. Mus. Zool., St. Petersburg 15: 1.
1930 *Allolobophora* (part.)–Stephenson, Oligochaeta, p. 905, 906, 907, 908.
1941 *Allolobophora* (part.)–Pop, Zool. Jb. Syst. 74: 20.
1956 *Allolobophora* (part.)–Omodeo, Arch. Zool. It. 41: 180.
1972 *Allolobophora* (part.)–Gates, Trans. Amer. Philos. Soc. 62(7): 68, 69.
1972 *Allolobophora*–Bouché, Inst. Natn. Rech. Agron., p. 263, 417.
1975 *Allolobophora*–Gates, Megadrilogica 2(1): 3.

Type Species
Lumbricus riparius Hoffmeister, 1843 (= *Enterion chloroticum* Savigny, 1826)

Diagnosis
Calciferous gland, opening into gut through a pair of vertical sacs posteriorly in x. Calciferous lamellae continued along lateral walls of sacs. Gizzard, mostly in xvii. Extraoesophageal vessels, passing to dorsal trunk in xii. Hearts, in vi–xi. Nephridial bladders, *J*-shaped, closed end laterally, ducts passing into parieties near *B*. Nephropores, inconspicuous, behind the clitellum irregularly alternating between levels slightly above *B* and above *D*. Setae paired. Prostomium, epilobic. Longitudinal musculature, pinnate. Colour variable. (after Gates, 1972c: 68 and 1975a: 3).

Discussion
Allolobophora was erected by Eisen (1873) without the designation of a type species and this situation was not corrected by Michaelsen (1900b) in his revision of the Lumbricidae. Typification of the genus was by Omodeo (1956) who selected *A. chlorotica* to be the type. Additional species that Eisen included in his *Allolobophora* were: *arborea, foetida, mucosa, norvegica, subrubicunda*, and *turgida*, none of which is now referable to this genus (Gates, 1975b: 7).

Allolobophora chlorotica (Savigny, 1826)

Green worm Ver vert

(Fig. 4)

1826 *Enterion chloroticum* + *E. virescens* Savigny, Mém. Acad. Sci. Inst. Fr. 5: 183.

1828 *Lumbricus anatomicus* Dugès, Ann. Sci. Nat. 15(1): 289.

1837 *Lumbricus chloroticus*–Dugès, Ann. Sci. Nat., ser. 2, 8: 17, 19.

1843 *Lumbricus riparius* Hoffmeister, Arch. Naturg. 9(1): 189.

1845 *Lumbricus communis luteus* Hoffmeister, Regenwürmer, p. 29.

1865 *Lumbricus viridis* Johnston, Cat. British non-paras. worms, p. 60.

1873 *Allolobophora riparia*–Eisen + *A. mucosa* Eisen, Öfv. Vet.-Akad. Förh. Stockholm 30(8): 46, 47.

1882 *Allolobophora neglecta* Rosa, Atti Acc. Torino 18: 170.

1884 *Allolobophora chlorotica*–Vejdovský, Syst. Morph. Oligo., p. 60.

1885 *Aporrectodea chlorotica*–Örley, Ertek. Term. Magyar Akad. 15(18): 22.

1892 *Allolobophora cambrica* Friend, Essex Nat. 6: 31.

1896 *Allolobophora curiosa* Ribaucourt + *A. Waldensis* Ribaucourt + *A. morganensis* Ribaucourt + *A. cambria* (laps.)–Ribaucourt, Rev. Suisse Zool. 4: 46, 47, 94.

1900 *Helodrilus (Allolobophora) chloroticus*–Michaelsen, Das Tierreich, Oligochaeta 10: 486.

Diagnosis

Length 30–70 mm, diameter 3–5 mm, segment number 80–138, prostomium epilobic, first dorsal pore 4/5. Clitellum xxviii, xxix–xxxvii. Tubercula pubertatis small, sucker-like discs on xxxi, xxxiii and xxxv. Setae closely paired, $AA > BC$, $DD = \frac{1}{2}$ C anteriorly, and $DD < \frac{1}{2}$ C posteriorly. Setae *c* and *d* on x often on white genital tumescences. Male pores in xv with large elevated glandular papillae extending over xiv and xvi. Seminal vesicles, four pairs in 9–12. Spermathecae, three pairs opening on level *cd* in 8/9, 9/10 and 10/11. Colour variable, frequently green but sometimes yellow, pink, or grey. Body cylindrical.

Biology

This species has been found in a wide variety of soil types, with a pH of 4.5–8.0, including gardens, fields, pastures, forests, clay and peat soils, lake shores and stream banks, estuarine flats, and among all sorts of organic debris. It has been found in caves in Europe and North America and also in botanical gardens and greenhouses in these same continents. Most of the sites where *A. chlorotica* was obtained in the present survey were moist, low areas, such as under various forms of debris and logs in ditches, relatively close to the Great Lakes. Eaton (1942) reported the habitat preference of this species as "wet and usually highly organic or polluted soil." In eastern Tennessee almost 85% of the specimens collected were from wet, highly organic habitats (Reynolds et al., 1974).

In appropriate conditions activity, including breeding, possibly occurs all year. In the northern part of the range there may be a single period of activity in the summer. There are records of active specimens occurring 300 mm below the soil surface although the species generally is characterized as shallow burrow-

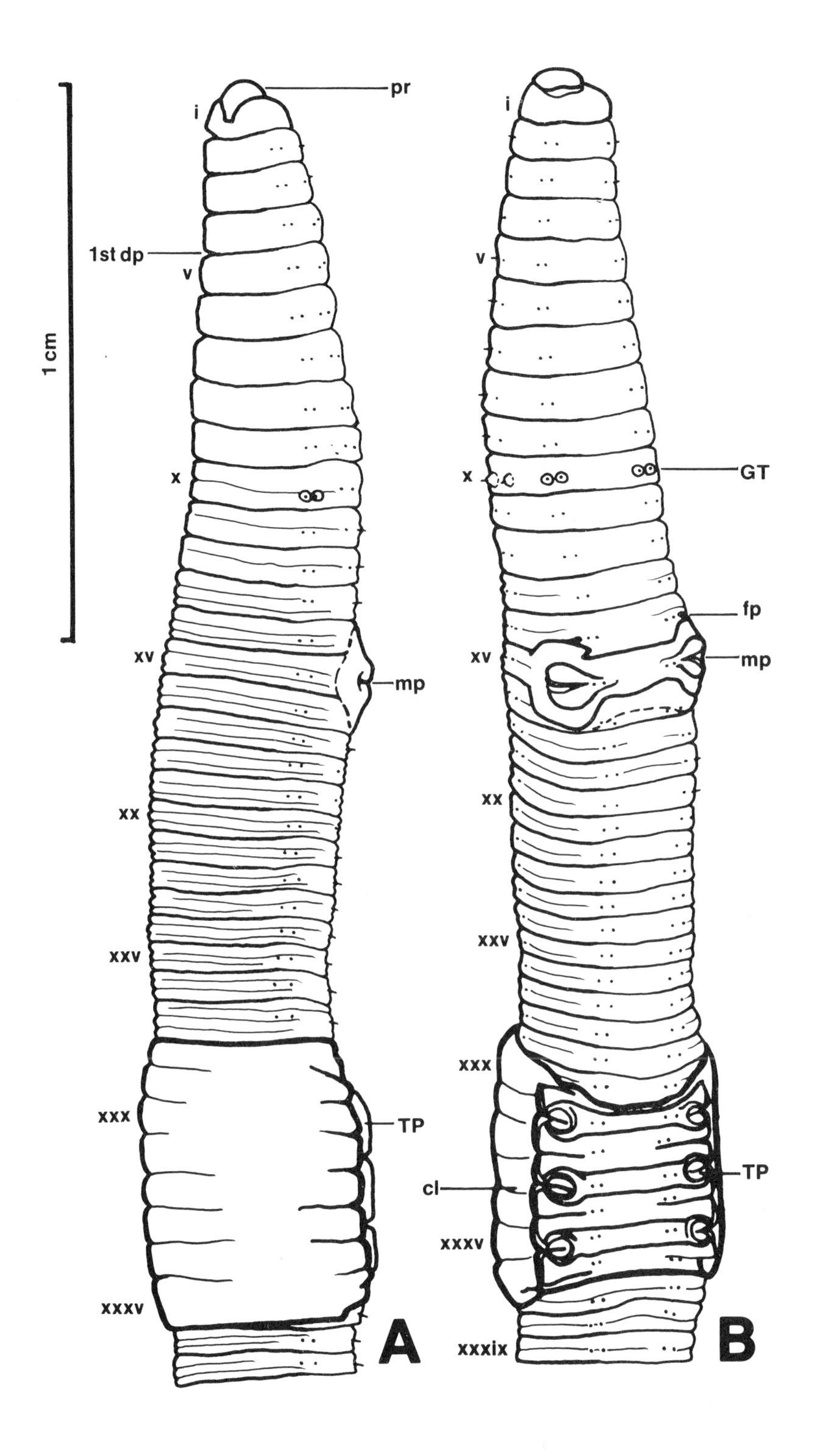

Fig. 4 External longitudinal views of *Allolobophora chlorotica* showing taxonomic characters. A. Dorsolateral view. B. Ventrolateral view. (ONT: Haldimand Co., cat. no. 7341)

ing. Defecation occurs below the soil surface as does copulation. *A. chlorotica* is obligatorily amphimictic (Reynolds, 1974c).

This species has been reported as the secondarily preferred host of the cluster fly, *Pollenia rudis* (Fabr.) (Yahnke and George, 1972; Thomson and Davies, 1973b); otherwise it is of minimal economic importance. It seems not to be preferred by fish, and anglers have found little use for it as bait.

Range

A native of Palaearctis, *A. chlorotica* is known from Europe, Iran, North America, South America, North Africa, and New Zealand (Gates, 1972c).

North American Distribution

British Columbia (Smith, 1917), New Brunswick (Reynolds, 1976d), Nova Scotia (Reynolds, 1975a, 1976a), Ontario (Reynolds, 1972a), Québec (Reynolds, 1975b, d, e, 1976c), Arizona (Gates, 1967), California (Smith, 1917), Connecticut (Reynolds, 1973c), Delaware (Reynolds, 1973a), District of Columbia (Smith, 1917), Idaho (Gates, 1967), Illinois (Smith, 1928), Indiana (Smith, 1917), Maine (Gates, 1961), Maryland (Reynolds, 1974b), Massachusetts (Reynolds, 1977), Michigan (Murchie, 1956), Missouri (Olson, 1936), Montana (Reynolds, 1972c), Nevada (Gates, 1967), New York (Olson, 1940), North Carolina (Smith, 1917), Ohio (Olson, 1928), Oregon (MacNab and McKey-Fender, 1947), Pennsylvania (Eaton, 1942), Tennessee (Reynolds, 1972b), Utah (Gates, 1967), Vermont (Gates, 1972c), Virginia (Gates, 1949), Washington (MacNab and McKey-Fender, 1947), West Virginia (Williams, 1942), Wisconsin (Gates, 1972c), Greenland (Levinsen, 1884). New records: Manitoba, Alaska, Minnesota.

Ontario Distribution (Fig. 5)

CARLETON CO. Reynolds (1972a); *Ottawa-Carleton Rd 34, Cumberland, under paper, 11 May 72, JWR, 0-0-1. DUNDAS CO. *Hwy 2, 5 km w of Iroquois, under logs, 11 May 72, JWR, 0-1-0. *Hwy 43, 1.29 km e of Chesterville, under rocks in ditch, 11 May 72, JWR, 0-0-4. *Hwy 31, 2.1 km s of Winchester Springs, under logs, 11 May 72, JWR, 0-0-3. DURHAM CO. *Hwy 401, .16 km w of Liberty Rd, Bowmanville, digging under log by railroad tracks, 15 May 72, JWR, 1-0-1. ELGIN CO. *Hwy 73, 3.06 km s of Harriettsville, under rotten log, 4 May 72, JWR, 0-1-1. *Hwy 73, 6.77 km n of Aylmer, under logs, 4 May 72, JWR, 3-2-4. *Hwy 3, 6.77 km e of Wallacetown, under pine logs, 4 May 72, JWR, 0-1-0. Hwy 3, 7.9 km w of Frome, under logs, 4 May 72, JWR, 0-0-7. ESSEX CO. *Hwy 3, 1.45 km e of Cottam, under logs, 4 May 72, JWR, 1-1-8. FRONTENAC CO. *Hwy 2, 1.13 km e of Westbrook, under logs and paper, 16 May 72, JWR, 0-0-10. *Hwy 15, 4.03 km n of Hwy 401, under paper in wet ditch, 16 May 72, JWR, 0-0-1. GRENVILLE CO. *Hwy 2, Johnstown, w.e., under logs, 11 May 72, JWR, 0-1-4. HALDIMAND CO. *Hwy 54, 2.26 km n of Caledonia, digging, 3 May 72, JWR, 2-2-0. HALTON CO. *Hwy 7, 4.84 km e of Georgetown, under burnt logs in soil, 4 May 73, JWR, 0-0-2. HASTINGS CO. *Hwy Jct 2 and 49, wet ditch, 15 May 72, JWR, 0-2-37. HURON CO. *Hwy 4, 1.29 km s of Wingham, under logs in wet area, 5 May 72, JWR, 0-5-7. KENT CO. *River Rd, 6.45 km w of Chatham, under logs, 23 Apr 72, JWR & TW, 0-1-0. Hwy 3, 1.77 km w of Palmyra, under logs, 4 May 72, JWR, 1-2-15. LAMBTON CO. Hwy 21, 2.1 km n of Wyoming, under logs in wet ditch, 4 May 72, JWR, 0-0-10. LANARK CO. *Hwy 43, 3.71 km e of Smiths Falls, under telephone pole in ditch, 11 May 72, JWR, 0-0-18. LEEDS CO. *Hwy 15, 2.26 km n of Seeley's Bay, under logs and concrete blocks in ditch, 16 May 72, JWR, 10-2-15. LENNOX AND ADDINGTON CO. *Hwy 2, 9.68 km w of Napanee, under paper in wet ditch, 15 May 72, JWR, 8-2-17. MIDDLESEX CO. Judd (1970). *Hwy 7, 2.26 km s of Parkhill, under posts, 4 May 72, JWR, 1-1-1. NIAGARA CO. *Niagara Co Rd 12, 3.06 km s of Grimsby, under paper in wet ditch, 1 May 72, JWR, 0-1-9. NIPISSING DIST. *Hwy 17. 1.29 km e of Verner, under paper in wet ditch, 13 May 72, JWR & JEM, 1-1-2. NORTHUMBERLAND CO. *Northumberland-Durham Rd 1, .65 km w of Hwy 33, under logs, 15 May 72, JWR, 3-0-3. ONTARIO CO. *Hwy 7, 1.61 km e of Green

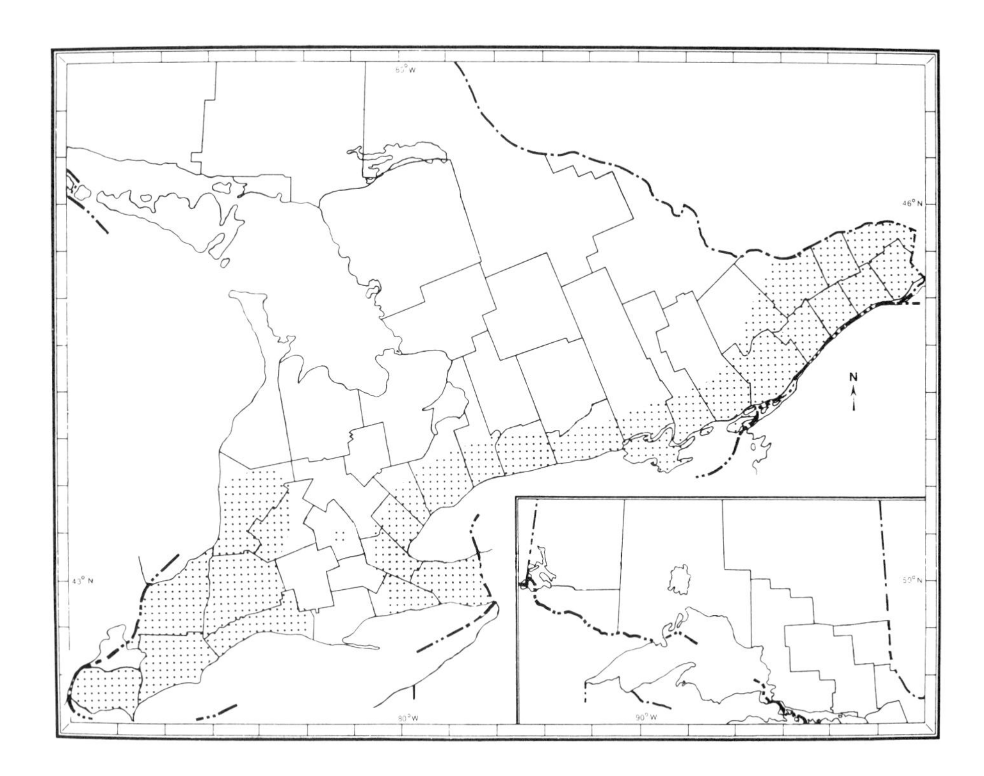

Fig. 5 The known Ontario distribution of *Allolobophora chlorotica*.

River, under logs, 26 Apr 72, JWR, 2-0-9. PEEL CO. *Hwy 5, .81 km e of Dixie Rd, 29 Apr 72, JWR, 0-0-2. *Hwy 7, Brampton, e.e., under debris, 4 May 73, JWR, 0-0-2. Hwy 401, 1.3 km w of Dixie Rd, wet ditch, 4 May 73, JWR, 0-0-1. PERTH CO. *Mitchell, next to Collegiate, under logs, 4 May 73, JWR & DWR, 2-3-2. PRESCOTT CO. *Hwy 34, 2.58 km n of Vankleek Hill, under logs and rocks, 11 May 72, JWR, 1-2-6. PRINCE EDWARD CO. *Hwy 33, 1.61 km s of Carrying Place, under log in wet ditch, 15 May 72, JWR, 3-0-8. *Hwy 33, Consecon, n.e., dump, 15 May 72, JWR, 0-0-1. *Hwy 33, 2.58 km e of Hillier, digging, 15 May 72, JWR, 0-3-0. Indian Point, in grab sample, approximately 1 metre depth in Lake Ontario, 24 Jul 74, DRB, 1-0-2, UW-0001. STORMONT CO. *Hwy 43, 5.48 km w of Finch, in and under wet straw in wet ditch, 11 May 72, JWR, 0-0-42. WATERLOO CO. Waterloo, Amos Ave., 1 Sep 75, DPS, 1-3-0, UW-0003. WENTWORTH CO. Reynolds (1972a). *Hwy 5, Waterdown, e.e, edge of corn (*Zea maize* L.) field, 29 Apr 72, JWR, 0-0-4. YORK CO. *Edenbrook Park, Islington, under logs and rocks near stream bank, 30 Apr 72, JWR & DWR, 1-1-2. Scarborough, 21 Kingston Rd., small wooded valley, 30 Nov 41, JO, 0-0-1, ROM.

Genus *Aporrectodea* Örley, 1885

1885 *Aporrectodea* Örley, Ertek. Term. Magyar Akad. 15(18): 22.
1900 *Allolobophora* (part.)–Michaelsen, Das Tierreich, Oligochaeta 10: 480.
1930 *Allolobophora* (part.)–Stephenson, Oligochaeta, p. 905, 906, 907, 908.
1941 *Allolobophora* (part.)–Pop, Zool. Jb. Syst. 74: 20.
1956 *Allolobophora* (part.)–Omodeo, Arch. Zool. It. 41: 180.
1972 *Allolobophora* (part.)–Gates, Trans. Amer. Philos. Soc. 62(7): 68, 69.
1972 *Allolobophora*–Gates, Bull. Tall Timbers Res. Stn. 12: 2.
1972 *Nicodrilus* Bouché, Inst. Natn. Rech. Agron., p. 315.
1975 *Aporrectodea*–Gates, Megadrilogica 2(1): 4.

Type Species

Lumbricus trapezoides Dugès, 1828.

Diagnosis

Calciferous gland, opening into gut through a pair of vertical sacs equatorially in x. Calciferous lamellae continued onto posterior walls of sacs. Gizzard, mostly in xvii. Extraoesophageal vessels, passing to dorsal trunk in xii. Hearts, in vi–xi. Nephridial bladders, *U*-shaped, ducts passing into parieties near *B*. Nephropores, inconspicuous, irregularly alternating between levels slightly above *B* and above *D*. Setae, paired. Prostomium, epilobic. Longitudinal musculature, pinnate. Pigment, if present, not red (after Gates, 1975a: 4).

Discussion

This forgotten genus originally included *Enterion chloroticum* Savigny, 1826 and *Lumbricus trapezoides* Dugès, 1828. Since Omodeo (1956) designated the former as the type species of *Allolobophora*, the latter automatically becomes the type for *Aporrectodea*. Bouché (1972) erected a new genus *Nicodrilus* with *Enterion caliginosum* Savigny, 1826 as the type and included *Lumbricus trapezoides* Dugès, 1828 in this new genus. Since *Aporrectodea* is a valid and available genus, *Nicodrilus* must be considered the junior synonym of *Aporrectodea*.

Aporrectodea icterica (Savigny, 1826)

Mottled worm Ver marbré

(Fig. 6)

1826 *Enterion ictericum* Savigny, Mém. Acad. Sci. Inst. Fr. 5: 183.
1837 *Lumbricus ictericus*–Dugès, Ann. Sci. Nat., ser. 2, 8: 17.
1886 *Allolobophora icterica*–Rosa, Atti Ist. Veneto, ser. 6, 4: 685.
1900 *Helodrilus ictericus*–Michaelsen, Das Tierreich, Oligochaeta 10: 500.
1926 *Bimastus tenuis*–Pickford, Ann. Mag. Nat. Hist., ser. 9, 17: 96.
1938 *Eophila icterica*–Tétry, Contr. Étude Faune Est France, p. 269.
1972 *Allolobophora icterica*–Bouché, Inst. Natn. Rech. Agron., p. 273.
1976 *Aporrectodea icterica*–Reynolds, Megadrilogica 2(12): 3.

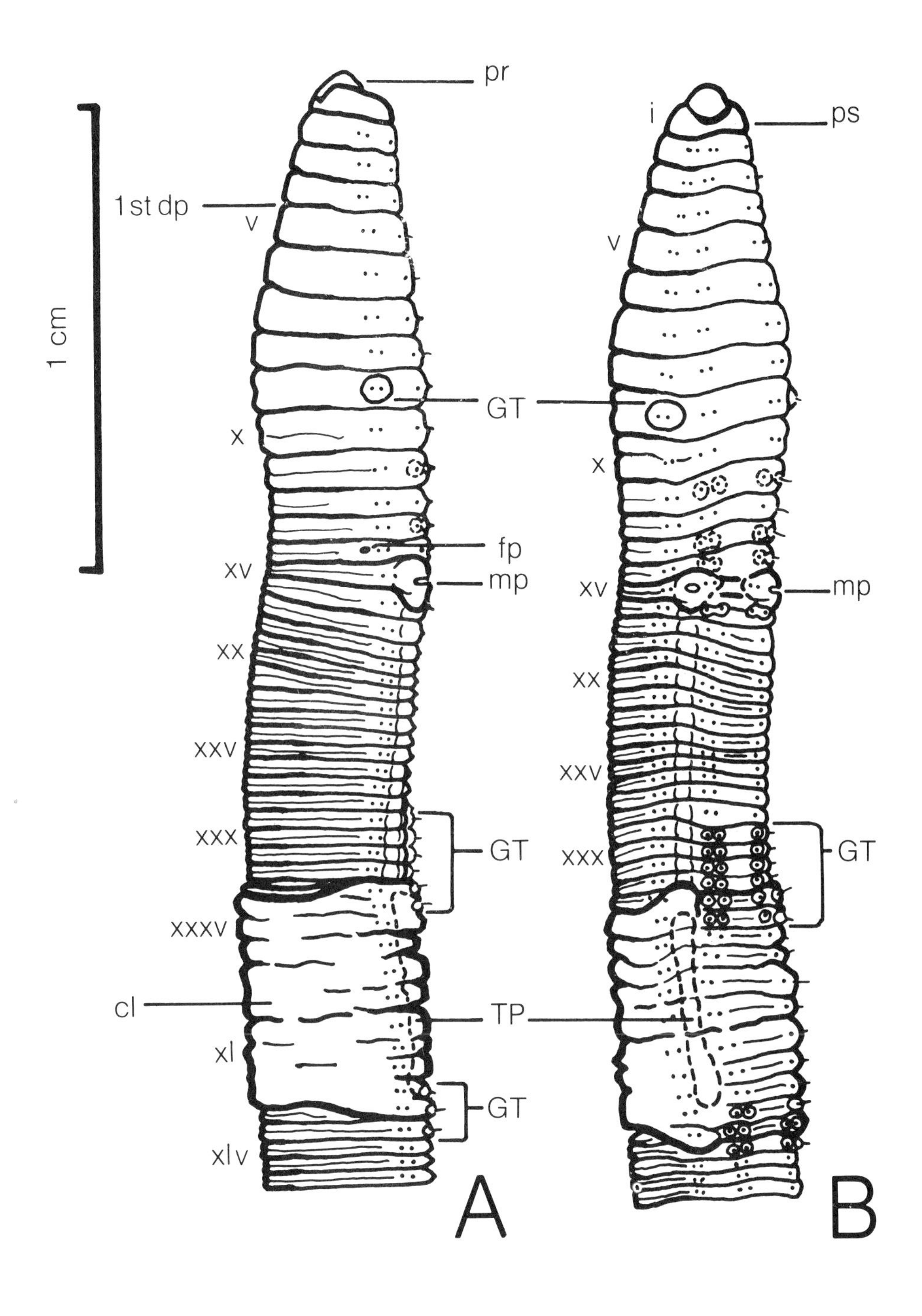

Fig. 6 External longitudinal views of *Aporrectodea icterica* showing taxonomic characters. A. Lateral view. B. Ventral view. (ONT: Wellington Co., cat. no. ROM I48)

Diagnosis
Length 55–135 mm, diameter 3–5 mm, segment number 140–190, prostomium epilobic, first dorsal pore 4/5. Clitellum xxxiii, xxxiv–xlii, xliii. Tubercula pubertatis in the form of a band xxxiv, xxxv–xli, xlii, xliii. Setae closely paired, posteriorly *AA*:*AB*:*BC*:*CD* = 45:5:25:4; *c* and *d* in form of genital tumescences in ix and *a* and *b* in xi–xvii, xxix–xxxiv, and xlii–xlv. Male pores on xv, minute, about at mid *BC*, with tumescences small and restricted to xv. Seminal vesicles, four pairs in 9–12, the anterior two pairs smaller. Spermathecae, three pairs with ducts opening on level *c* in 8/9–10/11, sometimes an anterior pair opening in 7/8. Colour, lacking. Body cylindrical.

Biology
In Europe Černosvitov and Evans (1947), Gerard (1964), and Tétry (1938) reported the species from garden soil, meadows, and orchards. With the exception of Bouché's (1972) study in France, *Ap. icterica* has been reported infrequently and in low numbers in Europe. The species is obligatorily amphimictic (Gates, 1968).

Ap. icterica is not known to have any economic importance.

Range
A native of Palaearctis, *Ap. icterica* is now known from western Europe and North America (Reynolds, 1976e; Schwert, 1977).

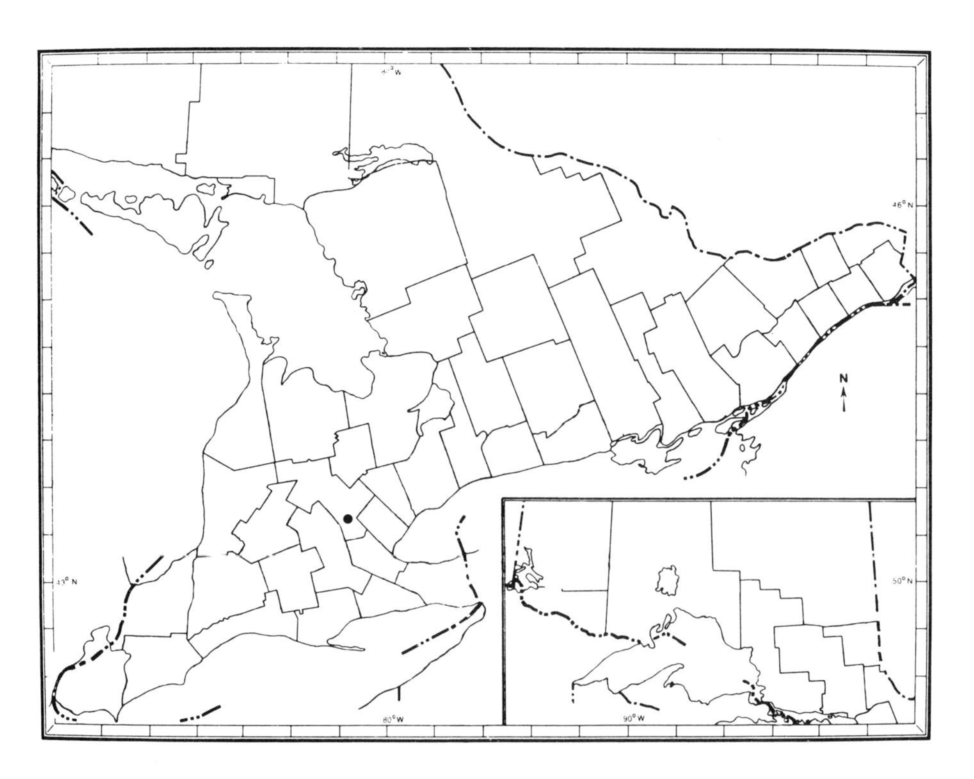

Fig. 7 The known Ontario distribution of *Aporrectodea icterica*.

North American Distribution
Ontario (Schwert, 1977), New York (Reynolds, 1976e).

Ontario Distribution (Fig. 7)
WELLINGTON CO. Schwert (1977). University of Guelph Arboretum, May-Jul 1976, DPS.

Aporrectodea longa (Ude, 1885)
Black head worm Ver à tête noire
(Fig. 8)

1826 *Enterion terrestre* (non 1820) Savigny, Mém. Acad. Sci. Inst. Fr. 5: 180.
1837 *Lumbricus terrestris*–Dugès, Ann. Sci. Nat., ser. 2, 8: 17, 18.
1845 *Lumbricus agricola* (non 1842) (part.) Hoffmeister, Regenwürmer, p. 5.
1885 *Allolobophora longa* Ude, Z. Wiss. Zool. 43: 136.
1889 *Lumbricus terrestris* + *L. longus*–L. Vaillant, Hist. Nat. Annel. 3(1): 113, 121.
1893 *Allolobophora terrestris*–Rosa, Mem. Acc. Torino, ser. 2, 43: 424, 444.
1900 *Helodrilus (Allolobophora) longus*–Michaelsen, Das Tierreich, Oligochaeta 10: 483.
1972 *Allolobophora longa*–Gates, Bull. Tall Timbers Res. Stn. 12: 114.
1972 *Nicodrilus longus longus*–Bouché, Inst. Natn. Rech. Agron., p. 322.
1975 *Aporrectodea longa*–Reynolds, Megadrilogica 2(4): 5.

Diagnosis
Length 90–150 mm, diameter 6–9 mm, segment number 150–222, prostomium, epilobic, first dorsal pore 12/13. Clitellum xxvii, xxviii–xxxiv, xxxv. Tubercula pubertatis xxxii–xxxiv. Setae closely paired, posteriorly $AA:AB:BC:CD = 60:7:28:5$; *a* and *b* in form of genital setae on genital tumescences in ix, x, xi, xxxi, xxxiii, xxxiv, and sometimes xii. Male pores on xv with elevated glandular borders, sometimes extending to xiv and xvi. Seminal vesicles, four pairs in 9–12, the anterior pairs smaller. Spermathecae, two pairs with short ducts opening on level *c* in 9/10 and 10/11. Colour, grey or brown with slight iridescence dorsally. Body cylindrical and dorsoventrally flattened posteriorly.

Biology
In Europe Černosvitov and Evans (1947) and Gerard (1964) reported the species from cultivated soil, gardens, pastures, and woodlands, and found it to be abundant in soils bordering rivers and lakes. According to Gates (1972c) *Ap. longa* is found in soils with a pH of 4.5 to 8.0, in greenhouses and botanical gardens, lawns, peat bog, in compost and under manure, including chicken yards and cow yards, and in many other types of soils. The species is known from caves in Europe.

In appropriate circumstances year-round activity is possible. Feeding, which takes place nocturnally on the surface of the soil, seems to be selective and leaves occasionally are dragged into the burrows. Casts are deposited on the surface and *Ap. longa*, together with *Ap. nocturna* (Evans, 1946), is believed re-

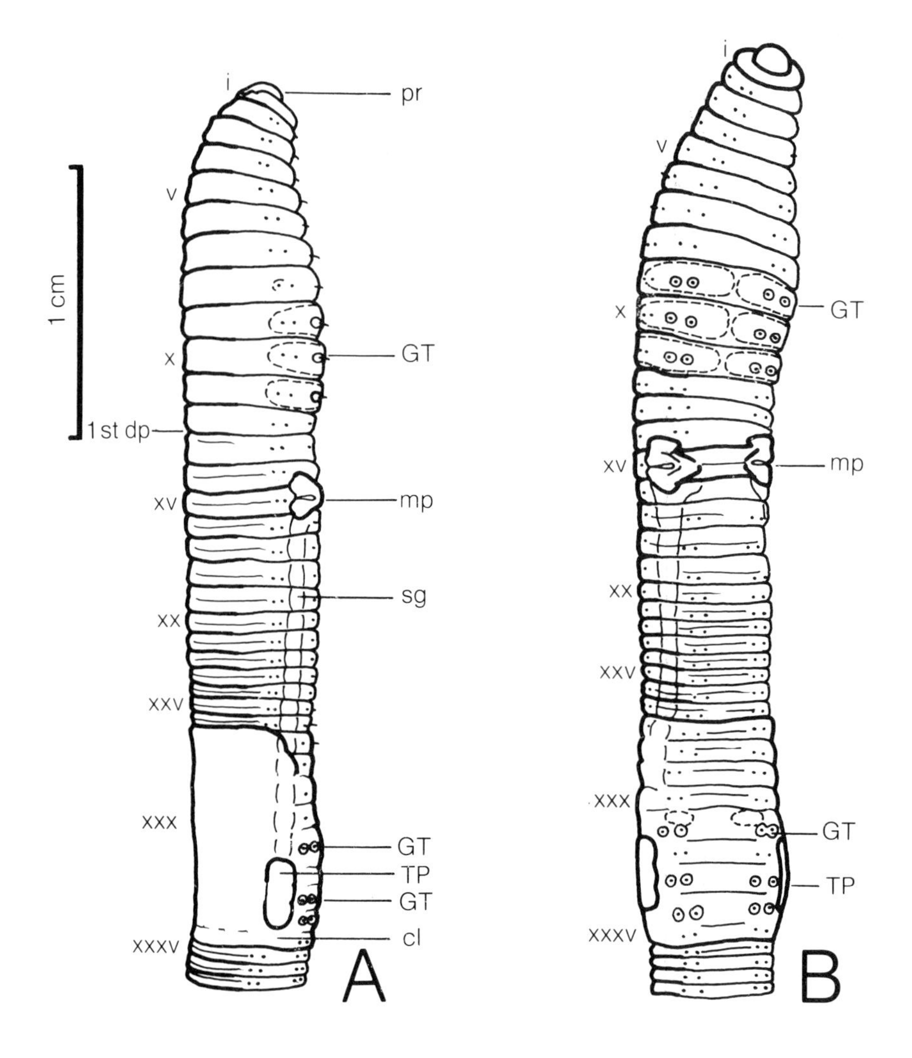

Fig. 8 External longitudinal views of *Aporrectodea longa* showing taxonomic characters. A. Lateral view. B. Ventral view. (NS: Cumberland Co., cat. no. 9206)

sponsible for the surface castings in England that Darwin studied so intensively. This species is obligatorily amphimictic with copulation beneath the soil surface (Gates, 1972a, 1972c, Reynolds, 1974c).

Ap. longa is not known to have been sold or used for bait in North America and is of little, if any, economic importance.

Range

A native of Palaearctis, *Ap. longa* is now known from Europe, North America, Central America, Africa, and Australasia (Gates, 1972c), and also now from Iceland (Gates 1972a).

North American Distribution

Owing to past confusion in the *Aporrectodea trapezoides* species group (cf. Gates, 1972a), many of the records of *Ap. longa* may have been attributed to other species such as *Lumbricus terrestris* or *Allolobophora terrestris* and various forms of varieties. There are a few reports of limited collections of this species in North America (cf. Eaton, 1942; Gates, 1953a; Murchie, 1956; Reynolds, 1973c, 1975a-c, 1976a, c; and Reynolds et al., 1974).

New Brunswick (Reynolds, 1976d), Nova Scotia (Reynolds, 1975a, 1976a), Ontario (Smith, 1917), Prince Edward Island (Reynolds, 1975c), Québec (Reynolds, 1975d, e, 1976c), Alabama (Gates, 1972a), California (Gates, 1972a), Colorado (Gates, 1967), Connecticut (Reynolds, 1973c), Indiana (Smith, 1917), Maine (Smith, 1917), Maryland (Reynolds, 1974b), Massachusetts (Reynolds, 1977), Michigan (Murchie, 1956), New Hampshire (Gates, 1972a), New Jersey (Gates, 1972a), New York (Eaton, 1942), North Carolina (Gates, 1972a), Ohio (Gates, 1972c), Oregon (MacNab and McKey-Fender, 1947), Pennsylvania (Bhatti, 1965), Tennessee (Reynolds, 1974a), Vermont (Gates, 1972a).

Ontario Distribution (Fig. 9)

Aporrectodea longa is reported here from Ontario for only the second time. The first report was made by Smith (1917). Both reports are from the greater Toronto area.

YORK CO. Smith (1917). *Edenbrook Park, Islington, under logs and rocks, 30 Apr 72, JWR and DWR, 0-1-3.

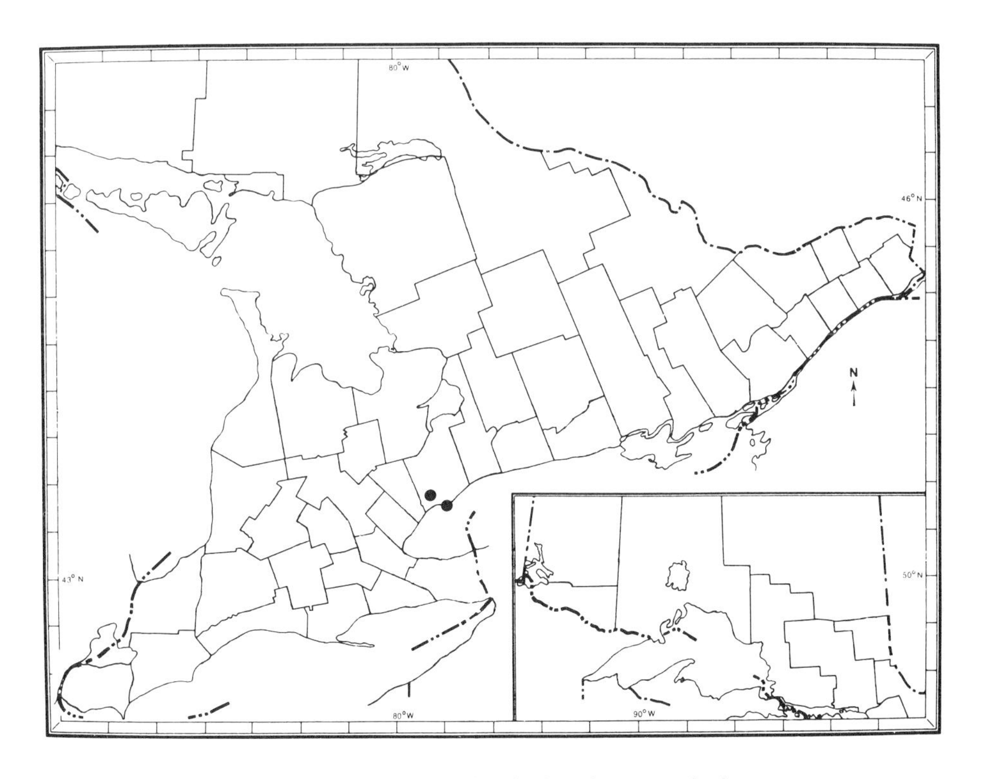

Fig. 9 The known Ontario distribution of *Aporrectodea longa*.

Aporrectodea trapezoides (Dugès, 1828)
Southern worm Ver méridional
(Fig. 10)

1828 *Lumbricus trapezoides* Dugès, Ann. Sci. Nat. 15(1): 289.
1917 *Helodrilus (Helodrilus) mariensis* Stephenson, Rec. Indian Mus. 13: 414.
1923 *Allolobophora (Eophila) mariensis*–Stephenson, Fauna British India, Oligochaeta, p. 504.
1931 *Allolobophora caliginosa trapezoides*–Chen, Cont. Biol. Lab. Sci. Soc. China (Zool.) 7(3): 168.
1941 *Allolobophora caliginosa* f. *trapezoides*–Gates, Proc. California Acad. Sci. (4), 23: 452.
1942 *Allolobophora caliginosa* (part.)–Eaton, J. Wash. Acad. Sci. 32(8): 246.
1948 *Allolobophora iowana* Evans, Ann. Mag. Nat. Hist., ser. 11, 14: 515.
1956 *Allolobophora (Microeophila) mariensis*–Omodeo, Arch. Zool. It. 41: 184.
1969 *Allolobophora longa* (part.)–Reinecke and Ryke, Rev. Écol. Biol. Sol. 6: 515.
1969 *Allolobophora caliginosa* (part.)–Støp-Bowitz, Nytt. Mag. Zool. 17(2): 191.
1972 *Nicodrilus (Nicodrilus) caliginosus meridionalis* Bouché, Inst. Natn. Rech. Agron., p. 334.
1972 *Allolobophora trapezoides*–Gates, Bull. Tall Timbers Res. Stn. 12:2.
1973 *Allolobophora caliginosa* f. *trapezoides*–Plisko, Fauna Polski, no. 1, p. 108.
1975 *Aporrectodea trapezoides*–Reynolds, Megadrilogica 2(3): 3.

Diagnosis

Length 80–140 mm, diameter 3–7 mm, segment number 93–169, generally > 130, prostomium epilobic, first dorsal pore 12/13, usually. Clitellum xxvii, xxviii–xxxiii, xxxiv. Tubercula pubertatis xxxi–xxxiii. Setae closely paired, posteriorly $AA > AB$, $DD < ½C$. Genital tumescences including *a* and *b* setae only, in ix–xi, xxxii–xxxiv, often in xxviii and occasionally in the region of xxvi–xxix. Male pores on xv. Seminal vesicles, four pairs in 9–12. Spermathecae, two pairs opening in 9/10 and 10/11. Colour variable and often lighter behind the clitellum until near the hind end, then deeper, slate, brown, brownish, reddish brown, and occasionally almost reddish, but not purple. Body dorsoventrally flattened posteriorly so that a cross section is nearly transversely rectangular with setal couples at the corners.

Biology

This species is found in a wide variety of habitats, according to the material examined by Gates (1972a). Similar statements have been made by Smith (1917), Olson (1928), Eaton (1942), Gates (1967), Reynolds (1973 a-c), and Reynolds et al. (1974). According to Gates (1972a, c) *Ap. trapezoides* is found in the earth around the roots of potted plants, in gardens, cultivated fields, forest soils of various types, on the banks of streams, and sometimes in sandy soil. It has been recorded from caves in North America and Afghanistan, and in California and Arizona may occur at elevations of 1525 m or more.

Activity may be year round under suitable conditions but it is not possible yet to make a similar statement concerning breeding. *Ap. trapezoides* is parthenogenetic, sometimes with pseudogamy, and male sterility is also common (Gates, 1972c; Reynolds, 1974c).

This species is often found in earthworm culture beds and is one of the five species commonly sold and used for bait in North America (Gates, 1972c).

It should be noted that many literature records of this species must be treated with caution because of taxonomic confusion. For a long time *Ap. trapezoides* was considered to be a variety of subspecies of *A. caliginosa* but it is unlikely that all references to *A. caliginosa* subspecies, variety, or forma, *trapezoides* do in fact refer to *Ap. trapezoides* (cf. Gates, 1972a: 4).

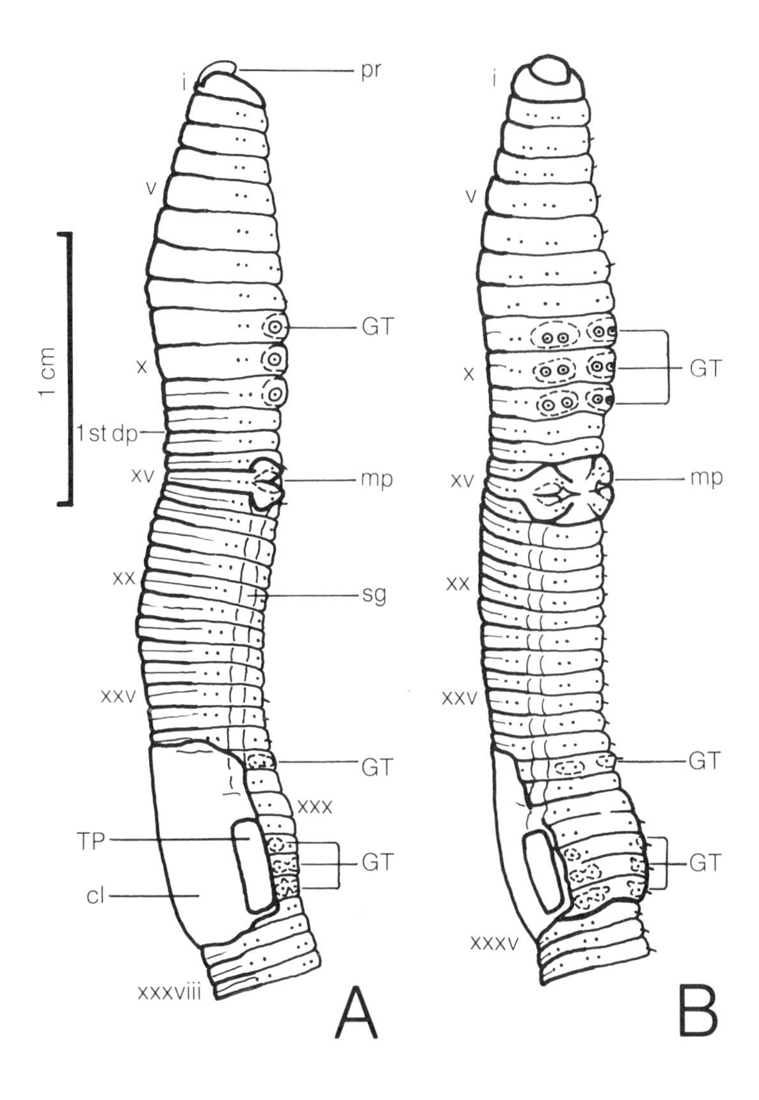

Fig. 10 External longitudinal views of *Aporrectodea trapezoides* showing taxonomic characters. A. Lateral view. B. Ventral view. (ONT: Waterloo Co., cat. no. 8002)

Range

A native of Palaearctis, *Ap. trapezoides* is now known from Europe, North America, South America, Africa, Asia, and Australasia (Gates, 1972c), and also from Iceland (Backlund, 1949).

North American Distribution

Alberta (Gates, 1972a), British Columbia (Gates, 1972a), Manitoba (Gates, 1972a), New Brunswick (Reynolds, 1976d), Nova Scotia (Reynolds, 1975a, 1976a), Ontario (Reynolds, 1972a), Prince Edward Island (Reynolds, 1975c), Québec (Reynolds, 1975b, e, 1976c), Alabama (Gates, 1972a), Alaska (Gates, 1972a) Arizona (Gates, 1972a), Arkansas (Gates, 1972a), California (Gates, 1967), Colorado (Gates, 1967), Connecticut (Reynolds, 1973c), Delaware (Reynolds, 1973a), District of Columbia (Gates, 1966), Florida (Gates, 1972a), Georgia (Gates, 1972a), Idaho (Gates, 1967), Illinois (Gates, 1972a), Indiana (Gates, 1972a), Iowa (Gates, 1967), Kentucky (Gates, 1959), Louisiana (Harman, 1952), Maine (Gates, 1966), Maryland (Reynolds, 1974b), Massachusetts (Reynolds, 1977), Michigan (Gates, 1972a), Minnesota (Gates, 1972a), Mississippi (Gates, 1972a), Missouri (Gates, 1967), Montana (Reynolds, 1972c), Nebraska (Michaelsen, 1910), Nevada (Gates, 1967), New Hampshire (Gates, 1972a), New Jersey (Gates, 1972a), New Mexico (Reynolds et al., 1974), New York (Gates, 1972a), North Carolina (Černosvitov, 1942), Ohio (Gates, 1972a), Oklahoma (Gates, 1967), Oregon (Gates, 1972a), Pennsylvania (Gates, 1972a), Rhode Island (Reynolds, 1973b), South Carolina (Gates, 1972a), Tennessee (Reynolds, 1974a), Texas (Reddell, 1965), Utah (Gates, 1967), Virginia (Gates, 1972a), Washington (Gates, 1972c), West Virginia (Gates, 1972a), Wisconsin (Gates, 1972c), Wyoming (Gates, 1967).

Ontario Distribution (Fig. 11)

Aporrectodea trapezoides was first reported from Ontario by Reynolds (1972a).

BRANT CO. *Hwy. 53, 2.74 km e of Mt. Vernon, under logs, 1 May 72, JWR, 0-0-3. *Hwy 54, 5.16 km e of Middleport, in ditch, 3May 72, JWR, 1-0-1. BRUCE CO. *Hwy 4, 6.77 km s of Teeswater, under logs, 5 May 72, JWR, 0-0-0-1. *Hwy 4, .48 km s of Hwy 9, under logs, 5 May 72, JWR, 2-4-4. Hwy 9, .81 km w of Bervie, under logs in pasture, 5 May 72, JWR, 0-0-1. *Hwy 21, 6.94 km n of Kincardine, under logs, 5 May 72, JWR, 0-3-0-1. Hwy 21, 2.76 km n of Port Elgin, under logs, 5 May 72, JWR, 6-0-2-1. CARLETON CO. *Ottawa-Carleton Rd 34, Cumberland, under paper, 11 May 72, JWR, 1-1-3. *Ottawa-Carleton Rd 34, .97 mi n of Leonard, under paper in wet ditch, 11 May 72, JWR, 2-0-4. DUFFERIN CO. *Hwy 9, 2.1 km w of Hwy 104, digging in road bank, 2 May 72, JWR, 0-0-1. *Hwy 9, 3.71 km w of Orangeville, under logs in ditch, 2 May 72, JWR, 0-0-1. DUNDAS CO. *Hwy 2, 5.48 km e of Iroquois, under logs, 11 May 72, JWR, 0-1-0. *Hwy 31, 2.1 km s of Winchester Springs, under logs, 11 May 72, JWR, 1-0-2. *Hwy 43, 1.29 km e of Chesterville, under rocks in ditch, 11 May 72, JWR, 3-0-2. *Hwy 43, 1.77 km e of Hallville, under logs, 11 May 72, JWR, 1-5-2. DURHAM CO. *Hwy 401, .16 km w of Liberty Rd, Bowmanville, digging under log by railroad tracks, 15 May 72, JWR, 0-0-4. ELGIN CO. *Hwy 73, 3.06 km s of Harrietsville, under rotten log, 4 May 72, JWR, 1-3-1. ESSEX CO. Hwy 3, 3.87 km w of Leamington, under lumber in dump, 4 May 72, JWR, 1-1-1. *Hwy 3, Ruthven, n.e., in wet ditch by railroad tracks, 4 May 72, JWR, 0-1-0. *Hwy 3, 1.45 km e of Cottam, under logs, 4 May 72, JWR, 1-0-11. FRONTENAC CO. *Hwy 15, 4.68 km s of Seeley's Bay, under logs, 16 May 72, JWR, 0-0-1. GLENGARRY CO. *Hwy 401, 7.42 km w of Summerstown Rd, under log, 11 May 72, JWR, 0-2-1. *Hwy 34, 2.58 km n of Lancaster, under dung in pasture, 11 May 72, JWR, 0-0-1. *Hwy 34, 11.13 km n of Lancaster, under log, 11 May 72, JWR, 4-1-4. GRENVILLE CO. Hwy 43, Merrickville, under logs, 11 May 72, JWR, 1-1-1. HALIBURTON CO. Reynolds (1972a). *Hwy 121, 3.22 km e of Tory Hill, dump, 16 May 72, JWR, 3-2-7. HALTON CO. *Hwy 7, 4.84 km e of Georgetown, under burnt logs in soil, 4 May 73, JWR, 0-0-1. Halton Co. Rd, 8.87 km n of Halton Co. Rd 8, wet ditch under logs by digging, 4 May 73, JWR, 0-2-0. HASTINGS CO. *Hwy Jct 62 and 620, under logs, 27 Apr 72, JWR, 0-0-1. HURON CO. *Hwy 4, 3.71 km s of Brucefield, under logs and rocks, 5 May 72, JWR, 0-1-0. Hwy 4, 5.16 km s of Clinton, under logs, 5 May 72, JWR, 0-0-2. *Hwy 4, 2.74 km n of Clinton, under logs, 5 May 72, JWR, 0-0-1. KENT CO. *Hwy 3, 3.55 km e of Wheatley, digging, 4 May 72, JWR, 5-0-0. LAMBTON CO. *Hwy 21, Edy's Mills, s.e., under railroad ties, 4 May 72, JWR, 0-0-1. LANARK CO. *Hwy 43, 5.32 km w of Smiths Falls, under logs, 11 May 72, JWR, 5-1-27. *Hwy 7, 5.32 km e of

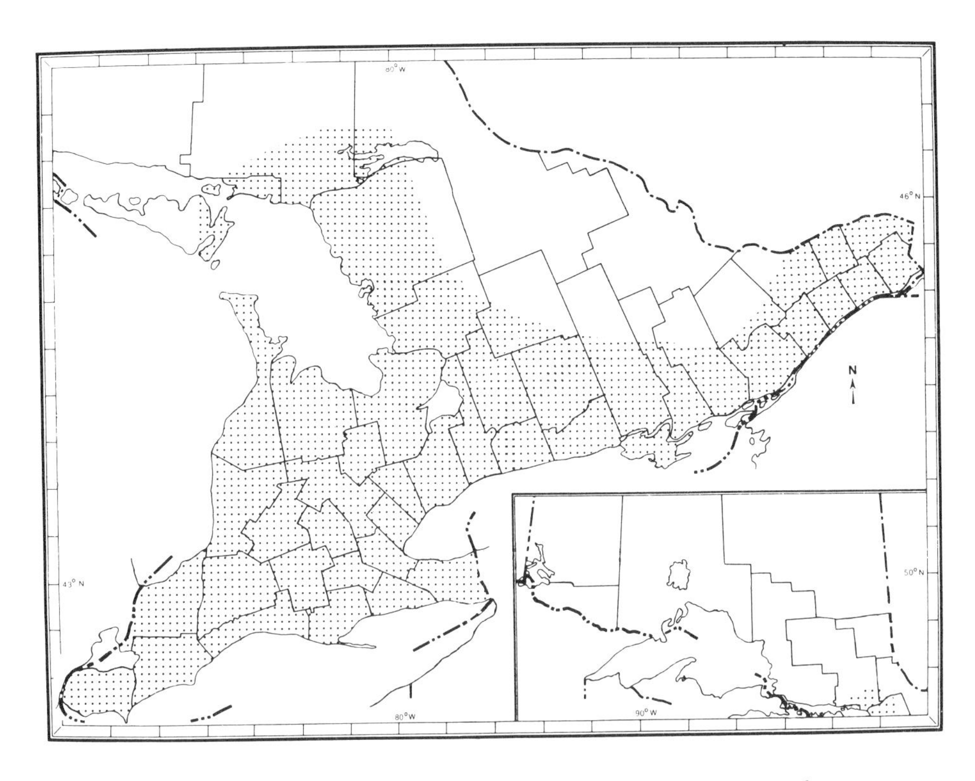

Fig. 11 The known Ontario distribution of *Aporrectodea trapezoides*.

Maberly, under dung in pasture, 11 May 72, JWR, 1-0-7. LEEDS CO. *Hwy 15, 3.22 km s of Morton, under log, 16 May 72, JWR, 0-0-2. *Hwy 15, 4.84 km n of Elgin, digging, 16 May 72, JWR, 0-0-7. LENNOX AND ADDINGTON CO. *Hwy 500, .81 km n of Denbigh, under logs, 16 May 72, JWR, 0-0-1. MANITOULIN DIST. *Hwy 68, 5.97 km n of Birch Island, under rocks, 13 May 72, JWR & JEM, 0-0-1. *Hwy 68, 3.06 km n of Birch Island, under logs and paper in ditch, 13 May 72, JWR & JEM, 1-2-1. *Hwy 68, Birch Island, s.e., under logs, 13 May 72, JWR & JEM, 1-2-4. *Hwy 68, 2.42 km s of Birch Island, under logs, 13 May 72, JWR & JEM, 0-0-3. MUSKOKA DIST. *Hwy 11, Melissa, s.e., under paper in ditch, 9 May 72, JWR, 0-0-1. NIAGARA CO. *Niagara Co. Rd 12, 3.06 km s of Grimsby, under paper in wet ditch, 1 May 72, JWR, 0-3-5. NIPISSING DIST. *Hwy 17, 8.39 km e of Warren, house site—digging, 13 May 72, JWR & JEM, 3-3-4. *Hwy 17, 5.48 km e of Warren, under logs and dung in pasture, 13 May 72, JWR & JEM, 0-0-1. ONTARIO CO. *Hwy 7, 1.61 km e of Green River, under logs, 26 Apr 72, JWR, 1-0-1. *Hwy 47, 10.48 km w of Uxbridge, digging, 9 May 72, JWR, 0-1-0. Dufferin Creek area, day after use of lampricide, 11 May 71, TY, 0-0-6, ROM-I40. OXFORD CO. *Hwy 59, 1.77 km e of Burgessville, under logs and wood chips, 1 May 72, JWR, 0-1-1. PARRY SOUND DIST. *Hwy 124, 3.06 km e of McKellar, under logs, 9 May 72, JWR, 0-1-3-1. PEEL CO. Hwy 401, 1.3 km w of Dixie Rd, wet ditch, 4 May 73, JWR, 1-1-6. PERTH CO. Hwy Jct 7, 8 and 59, Shakespeare, under logs, 3 May 72, JWR, 0-0-1. *Mitchell, next to Collegiate, under logs, 4 May 72, JWR & DWR, 0-0-1. PRESCOTT CO. *Hwy 34, 5.48 km s of Vankleek Hill, under logs, 11 May 72, JWR, 1-0-2. Hwy 17, 3.22 km w of Alfred, wet ditch, 11 May 72, JWR, 2-2-1. PRINCE EDWARD CO. *Hwy 33, Consecon, n.e., dump, 15 May 72, JWR, 0-1-4. *Hwy 49, 14.84 km n of Picton, under paper in ditch, 15 May 72, JWR, 0-1-4. RUSSELL CO. *Hwy 17, 4.35 km w of Rockland, under logs, 11 May 72, JWR, 1-1-0. SIMCOE CO. *Hwy 89, 1.94 km e of Rosemont, under logs in wet ditch, 2 May 72, JWR, 0-0-2. *Hwy 27, 5.97 km s of Cookstown, under logs, 2 May 72, JWR, 9-2-1. Hwy 27, 3.22 km s of Newton Robinson, wet ditch, 2 May 72, JWR, 0-1-1. Barrie, 14 Greenfield, under logs and digging in garden, 7 May 72, JWR & MKG, 4-1-0. *Barrie, park opposite 14 Greenfield, under rocks and grass clippings, 7 May 72, JWR & GWA, 0-

0-1. *Hwy 11, 15.16 km s of Severn Bridge, under logs, 9 May 72, JWR, 0-0-1. STORMONT CO. *Hwy 43, 5.48 km w of Finch, in and under straw in wet ditch, 11 May 72, JWR, 1-0-3. SUDBURY DIST. *Hwy Jct 64 and 69, under rocks, 13 May 72, JWR & JEM, 1-1-4. *Hwy 69, 1.94 km s of Hwy 607, under rocks, 13 May 72, JWR & JEM, 3-0-3. VICTORIA CO. *Hwy 35, 8.71 km s of Hwy 7, under log, 26 Apr 72, JWR, 0-0-2. *Hwy 7, 7.26 km e of Hwy 35, under log, 26 Apr 72, JWR, 0-1-1. *Hwy 7, 1.94 km w of Hwy 46, under logs and dung, 9 May 72, JWR, 0-1-2. WATERLOO CO. *Hwy 7 and 8, 2.58 km w of New Hamburg, under paper in wet ditch, 3 May 72, JWR, 4-1-8. Hwy 7 and 8, 1.77 km w of Petersburg, under log, 3 May 72, JWR, 1-0-4. Hwy 85, 1.45 km n of St. Jacobs, digging, 3 May 72, JWR, 0-0-3. Hwy 86, West Montrose, e.e., under logs, 3 May 72, JWR, 0-0-1. Waterloo, Beechwood South, on sidewalk after rain (evening), 15 Jun 75, DPS, 5-1-0-3, UW-0004. Waterloo, Amos Ave., 1 Sep 75, DPS, 0-0-0-1, UW-0003. WELLINGTON CO. *Hwy 24, 5.16 km e of Eramosa, under logs in cedar (*Thuja occidentalis*) woodlot, 29 Apr 72, JWR, 0-0-1. Hwy 6, University of Guelph, on wet driveway behind Soil Science Building, 2 May 72, JWR, 0-0-4. *Hwy 6, 10.64 km s of Arthur, under logs, 2 May 72, JWR, 2-1-0. WENTWORTH CO. *Hwy 5, Waterdown, e.e., edge of corn (*Zea maize*) field, 29 Apr 72, JWR, 0-0-1. YORK CO. *Hwy 27, 1.45 km n of Hwy 7, under logs, 29 Apr 72, JWR, 0-0-1. Edenbrook Park, Islington, quantitative study #2 (formalin), 18 May 72, JWR, 2-1-1. Toronto, 1875, GE, 0-0-4, USNM-4559. Toronto, 1875, GE, 0-0-2, USNM-19704. Scarborough Bluffs, Brimley Rd, garden, 28 Oct 41, JO, 0-0-1, ROM.

Aporrectodea tuberculata (Eisen, 1874)

Canadian worm Ver canadien

(Fig. 12)

1874 *Allolobophora turgida* f. *tuberculata* Eisen, Öfv. Vet-Akad. Förh. Stockholm 31(2): 43.

1910 *Allolobophora similis* Friend, Gardener's Chron. 48: 99.

1911 *Aporrectodea similis*–Friend, Zoologist, ser. 4, 15: 144.

1927 *Helodrilus caliginosus trapezoides*–Blake, Ill. Biol. Monogr. 10(4): 63.

1930 *Allolobophora turgida* (part.) + *Allolobophora trapezoides* (part.)–Bornebusch, Forstl. Forsøgs. Danm. 11: 94.

1942 *Allolobophora caliginosa* (part.)–Eaton, J. Wash. Acad. Sci. 32(8): 246.

1952 *Allolobophora arnoldi* Gates, Breviora, no. 9: 1.

1962 *Allolobophora arnoldi* + *A. nocturna*–Omodeo, Mem. Mus. Civ. Sto. Nat. Verona 10: 92.

1963 *Lumbricus terrestris*–Cameron and Fogal, Can. J. Zool. 41: 753.

1969 *Allolobophora caliginosa*–van Rhee, Pedobiologia 9: 130.

1969 *Allolobophora caliginosa* (part.)–Støp-Bowitz, Nytt. Mag. Zool. 17(2): 191.

1972 *Nicodrilus (Nicodrilus) caliginosus alternisetosus* Bouché, Inst. Natn. Rech. Agron., p. 333.

1972 *Allolobophora tuberculata*–Gates, Bull. Tall Timbers Res. Stn. 12: 44, 45.

1973 *Allolobophora caliginosa* f. *typica* (part.)–Plisko, Fauna Polski, no. 1, p. 107.

1975 *Aporrectodea tuberculata*–Reynolds, Megadrilogica 2(3): 3.

Diagnosis

Length 90–150 mm, diameter 4–8 mm, segment number 146–194, prostomium epilobic, first dorsal pore 11/12 or 12/13. Clitellum xxvii–xxxiv. Tubercula pubertatis xxx, xxxi–xxxiii, xxxiv. Setae closely paired, $AB \simeq CD$, $AA > BC$, $DD \simeq \frac{1}{2}C$. Genital tumescences absent in xxxi and xxxiii, present in xxx, xxxii, and xxxiv and frequently in xxvi. Male pores in xv between *b* and *c*. Seminal vesicles, four pairs in 9–12. Spermathecae, two pairs opening on level *c* in 9/10 and 10/11. Colour, unpigmented, almost white or greyish or sometimes with light pigmentation on the dorsum. Body cylindrical.

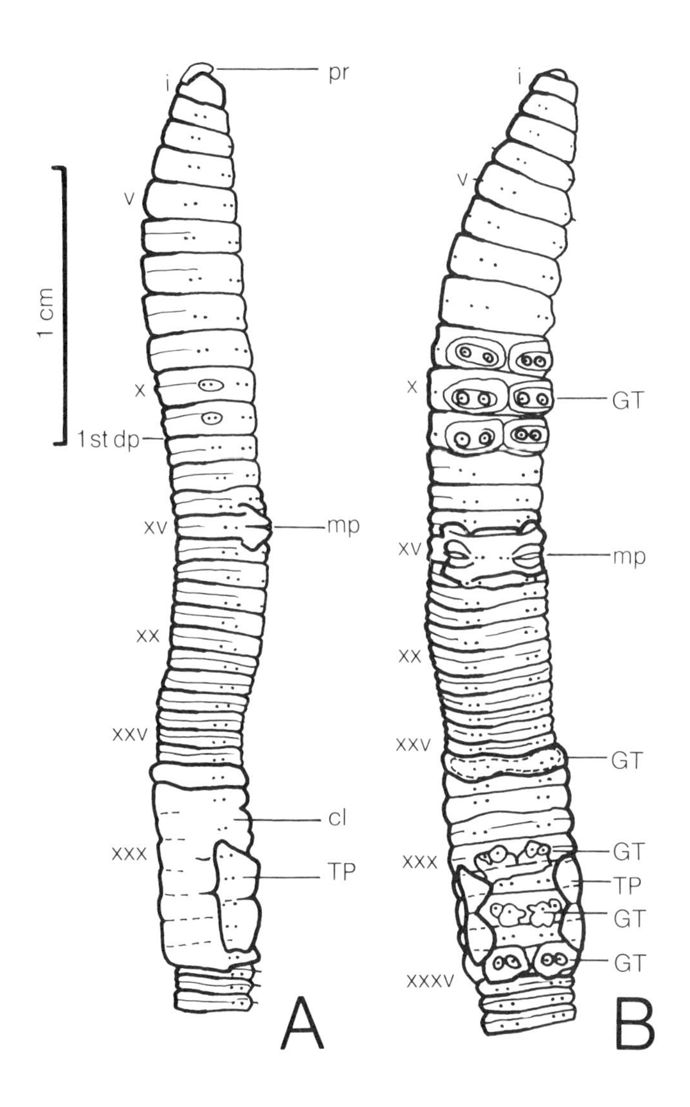

Fig. 12 External longitudinal views of *Aporrectodea tuberculata* showing taxonomic characters. A. Lateral view. B. Ventral view. (ONT: Sudbury Dist. cat. no. 8544)

Discussion

Aporrectodea tuberculata was described by Eisen in 1874 with specimens obtained from Niagara County (then Welland County), Ontario. Scandinavian specimens of *Ap. tuberculata* seem to have first been identified by Eisen as *Allolobophora cyanea* (fide Hoffmeister) (Gates, 1972a). Eisen recorded no Type Locality for *Ap. tuberculata*; but subsequently this has been designated as Niagara County, Ontario.

Biology

Gerard (1964) reported this species' habitat as pastureland. In the western United States Gates (1967) found it in wet areas near streams and springs where there was a large concentration of organic matter. In east Tennessee (Reynolds et al., 1974) *Ap. tuberculata* was recorded 75% of the time from ditches, or under logs or debris such as lumber. Gates (1972c) records it from soils of pH 4.8–7.5 including turf, compost, peat, bogs, and rarely manure. In Ontario this species was found primarily under logs and rocks in all but four eastern counties (cf. Fig. 11).

Under favourable conditions activity, including breeding, can be year round. However, Gates (1972c) states that probably throughout New England, New York, and Canada aestivation and hibernation are climatically imposed with breeding restricted to spring and late autumn. Reproduction in this species is obligatorily amphimictic with copulation beneath the soil surface (Reynolds, 1974c).

Ap. tuberculata has been obtained from culture beds on earthworm farms and being so common in gardens and fields is the species most likely to be dug for bait in much of Canada.

Range

A native of Palaearctis, *Ap. tuberculata* is now known from Europe, North America, South America, Asia, and Australia (Gates, 1972c), and also from Iceland (Backlund, 1949).

North American Distribution

Alberta (Gates, 1972a), British Columbia (Wickett, 1967), Manitoba (Gates, 1972a), New Brunswick (Reynolds, 1976d), Newfoundland (Gates, 1972a), Nova Scotia (Reynolds, 1975a, 1976a), Ontario (Eisen, 1874), Prince Edward Island (Reynolds, 1975c), Québec (Reynolds, 1975b, d, e, 1976c), Saskatchewan (Gates, 1972a), Alaska (Gates, 1972a), Arizona (Reynolds et al., 1974), California (Gates, 1967), Colorado (Gates, 1967), Connecticut (Reynolds, 1973c), Delaware (Reynolds, 1973a), Florida (Gates, 1972c), Idaho (Gates, 1967), Indiana (Gates, 1972a), Iowa (Gates, 1967), Maine (Gates, 1966), Maryland (Reynolds, 1974b), Massachusetts (Reynolds, 1977), Michigan (Gates, 1972a), Minnesota (Gates, 1972a), Montana (Reynolds, 1972c), Nevada (Gates, 1967), New Hampshire (Gates, 1972a), New Jersey (Gates, 1972a), New Mexico (Gates, 1967), New York (Gates, 1972a), North Carolina (Gates, 1972a), Ohio (Gates, 1972a), Oregon (Gates, 1972a), Pennsylvania (Gates, 1972a), Rhode Island (Reynolds, 1973a), Tennessee (Reynolds, 1974a), Utah (Gates, 1967), Vermont (Reynolds, 1976c), Virginia (Gates, 1972a), West Virginia (Gates, 1972a), Wisconsin (Gates, 1972a), Wyoming (Gates, 1967). New records: Arkansas, Illinois.

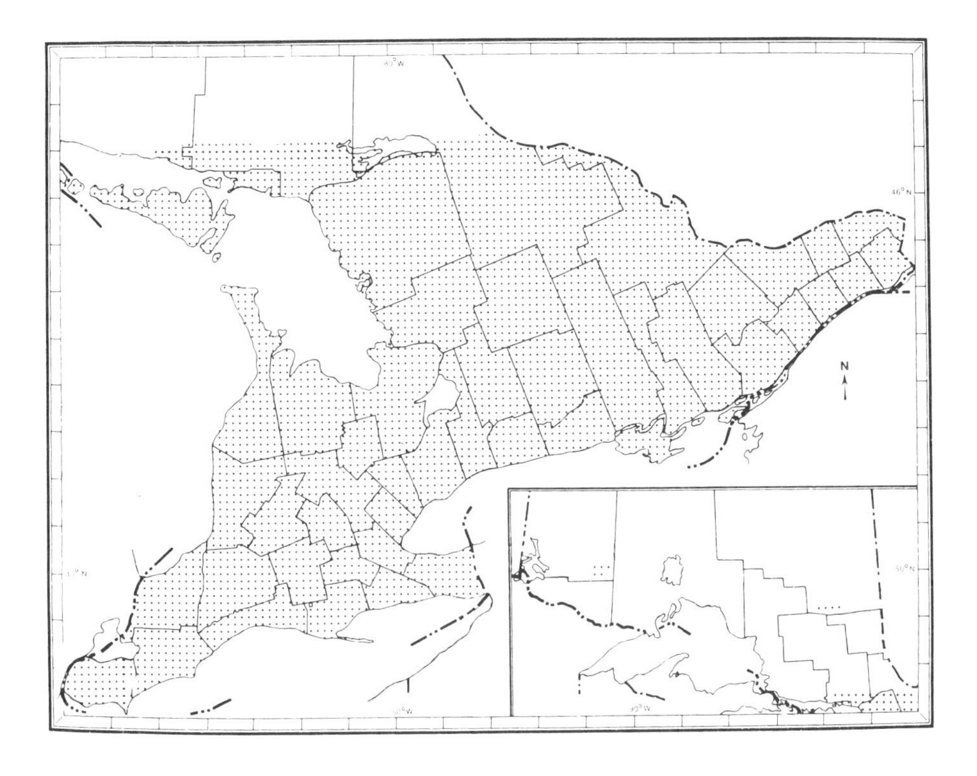

Fig. 13 The known Ontario distribution of *Aporrectodea tuberculata.*

Ontario Distribution (Fig. 13)

ALGOMA DIST. *Hwy 17, 7.58 km e of Spanish, under logs, 13 May 72, JWR & JEM, 4-0-2. *Hwy 17, .48 km e of Spanish, under logs and paper in ditch, 13 May 72, JWR & JEM, 22-16-13-1. Gargantua Bay, 11 Jun 75, DRB, 0-1-0-1, UW-0001. BRANT CO. *Hwy 53, 5.48 km w of Cathcart, under railroad ties, 1 May 72, JWR, 0-1-1. *Hwy 53, 4.35 km e of Cathcart, under logs, 1 May 72, JWR, 10-8-25. *Hwy 53, 2.74 km e of Mt. Vernon, under logs, 1 May 72, JWR, 1-0-7. *Hwy 54, 5.16 km e of Middleport, ditch, 3 May 72, JWR, 0-1-0. *Hwy 2, 5.45 km e of Paris, under lumber and paper, 3 May 72, JWR, 4-0-2. BRUCE CO. *Hwy 4, .48 km s of Hwy 9, under logs, 5 May 72, JWR, 0-0-2. *Hwy 21, 6.94 km n of Kincardine, under logs, 5 May 72, JWR, 0-2-6. Hwy 21, 2.76 km n of Port Elgin, under logs, 5 May 72, JWR, 0-0-4. *Hwy 21, 1.61 km e of Elsinore, under logs (pair *in copula*), 5 May 72, JWR, 4-4-2-1. COCHRANE DIST. Reynolds (1972a). DUFFERIN CO. Hwy 9, 1.61 km e of Hwy 10, under logs, 29 Apr 72, JWR, 17-1-0. *Hwy 9, 2.1 km w of Hwy 104, digging in road bank, 2 May 72, JWR, 4-1-13. *Hwy 9, 10.16 km w of Orangeville, under junk, 2 May 72, JWR, 0-0-1. *Hwy 9, 3.71 km w of Orangeville, under logs in ditch, 2 May 72, JWR, 6-4-10. *Hwy 10-24, 7.74 km n of Camilla, under logs, 2 May 72, JWR, 6-1-3. *Hwy 89, 10 km e of Primrose, wet ditch, 2 May 72, JWR, 0-0-10. DUNDAS CO. *Hwy 31, .81 km n of Hwy 43, under logs, 11 May 72, JWR, 0-0-3. DURHAM CO. *Hwy 7A, 11.77 km e of Hwy 7-12, digging in ditch, 26 Apr 72, JWR, 1. *Hwy 7A, 2.1 km e of Nestleton Station, under logs, 26 Apr 72, JWR, 2-7-3. *Hwy 401, .16 km of Liberty Rd, Bowmanville, digging under log next to railroad tracks, 15 May 72, JWR, 0-0-3. * 401, .32 km e of Mill St, Newcastle, under logs, 15 May 72, JWR, 0-0-5. *Hwy 401, 1.61 km Newtonville, under logs, 15 May 72, JWR, 2-1-8. Hwy 401, 1.61 km w of Hwy 28, under paper, May 72, JWR, 2-0-3. ELGIN CO. *Hwy 73, 3.06 km s of Harriettsville, under rotten log, 4 Ma, JWR, 0-3-9. Hwy 3, 9.19 km e of St. Thomas, under logs, 4 May 72, JWR, 9-3-0. *Hwy 3, 6. n e of Wallacetown, under pine logs, 4 May 72, JWR, 13-2-1. *Hwy 3, 2.1 km e of Wallacetown, under logs, 4 May 72, JWR, 1-0-4. Hwy 3, 3.87 km w of New Glasgow, under logs in abandoned school yard, 4 May 72, JWR, 9-4-3-1. ESSEX CO. *Hwy 3, 4.03 km e of Leamington, under logs, 4 May 72, JWR, 2-2-3. Hwy 3, 3.87 km w of Leamington, under lumber in dump, 4 May 72, JWR, 5-4-3. *Hwy 3, Ruthven, n.e., wet ditch by railroad tracks, 4 May 72, JWR, 4-1-0. FRONTENAC CO. Reynolds (1972a). *Hwy 7, 3.71 km e of Hwy 38, under log, 11 May 72, JWR, 4-0-6. *Hwy 2, .97 km w of Westbrook, under logs, 16 May 72, JWR, 0-0-2. *Hwy 2, 1.13 km e of Westbrook, under logs and paper, 16 May 72, JWR, 0-1-3. *Hwy 15, 4.68 km s of Seeley's Bay, under logs, 16 May 72, JWR, 5-1-1. *Hwy 7, 15.81 km e of Kaladar, under logs, 16 May 72, JWR, 2-0-3. GRENVILLE CO. *Hwy 2, 5.16 km w of Prescott, under rocks, 10 May 72, JWR, 3-2-1. Hwy 2, Cardinal, w.e., under log and paper, 11 May 72, JWR, 7-5-0. *Hwy 43, 3.87 km e of Kemptville, under logs, 11 May 72, JWR, 0-0-11. GREY CO. *Hwy 21, 7.26 km w of Springmount, under rock, 5 May 72, JWR, 0-0-1. *Hwy 6BP, Rockford, w.e., under logs, 5 May 72, JWR, 11-1-2. *Hwy 6, 3.06 km s of Chatsworth, under logs, 5 May 72, JWR, 9-2-10. *Hwy 6, 6.94 km s of Dornoch, under logs, 5 May 72, JWR, 1-1-2. *Hwy 10, 8.06 km s of Flesherton, under logs, 5 May 72, JWR, 5-1-1. HALDIMAND CO. *Hwy 3, 6.29 km w of Dunnville, under logs, 1 May 72, JWR, 5-1-2. *Hwy 3, 6.13 km w of Cayuga, under logs, 1 May 72, JWR, 0-1-5. *Hwy 54, 2.26 km n of Caledonia, digging, 3 May 72, JWR, 0-1-2. HALIBURTON CO. *Reynolds (1972a). HALTON CO. *Hwy 5, 9.19 km w of Hwy 10, under log next to barn, 29 Apr 72, JWR, 0-0-1. *Halton Co. Rd 3, Esquesing Community, under rocks and logs next to Town Hall, 4 May 73, JWR, 7-3-0. *Hwy 7, 4.84 km e of Georgetown, under burnt logs in soil, 4 May 73, JWR, 0-0-3. Halton Co. Rd 3, 8.87 km n of Halton Co. Rd 8, wet ditch under logs and digging, 4 May 73, JWR, 7-1-15. Halton Co. Rd 3, .48 km n of Halton Co. Rd 8, under logs in field, 4 May 73, JWR, 5-5-6-1. Rattlesnake Point, 1 Nov 41, JO, 0-0-2, ROM. HASTINGS CO. *Hwy 620, 5.16 km e of Coe Hill, under logs, 27 Apr 72, JWR, 1-0-1. *Hwy Jct 62 and 620, under logs, 27 Apr 72, JWR, 5-1-1. Hwy 62, 7.42 km n of Bannockburn, under logs, 27 Apr 72, JWR, 6-0-3. *Hwy 62, 1.61 km n of Hwy 7, under log in wet ditch, 27 Apr 72, JWR, 0-0-1. HURON CO. *Hwy 4, 3.71 km s of Brucefield, under log, 5 May 72, JWR, 0-0-1. *Hwy 4, 1.29 km s of Wingham, under logs in wet area, 5 May 72, JWR, 3-0-1. KENORA DIST. Rushing River Provincial Park, stream trickle in campground, 13-14 Jun 71, ROM Field Party, 0-1-0, ROM-I37. KENT CO. *River Rd, 6.45 km w of Chatham, under logs, 23 Apr 72, JWR & TW, 0-1-1. *Hwy 3, 9.03 km e of Port Alma, under log, 4 May 72, JWR, 9-0-4. LAMBTON CO. *Hwy 21, 6.45 km n of Dresden, under logs, 4 May 72, JWR, 3-6-7. *Hwy 21, Edy's Mills, s.e., under railroad ties, 4 May 72, JWR, 0-0-6. Hwy 21, 2.1 km n of Wyoming, under log in wet ditch, 5 May 72, JWR, 0-0-1. *Hwy 7, 1.61 km n of Arkona, under log, 4 May 72, JWR, 6-4-1. LANARK CO. *Hwy 43, 3.71 km e of Smiths Falls, under telephone pole in ditch, 11 May 72, JWR, 4-3-4. *Hwy 43, 5.32 km w of Smiths Falls, under logs, 11 May 72, JWR, 0-0-13. *Hwy 7, 4.35 km w of Perth, digging under gate post, 11 May 72, JWR, 3-0-12. LEEDS CO. *Hwy 15, 3.22 km s of Morton, under log, 16 May 72, JWR, 2-0-3. *Hwy 15, 5.64 km n of Morton, wet ditch, 16 May 72, JWR, 0-0-11. *Hwy 15, 4.84 km n of Elgin, digging, 16 May 72, JWR, 0-0-13. *Hwy 15, 4.19 km s of Lombardy, under logs and crawling on the surface during rain, 16 May 72, JWR, 11-14-28. *Hwy 15, 7.1 km n of Lombardy, under paper in ditch, 16 May 72, JWR, 0-0-1. LENNOX AND ADDINGTON CO. *Hwy 2, Napanee, w.e., under rocks behind EEEE Motel, 15 May 72, JWR, 2-1-2. *Hwy 2, 6.61 km w of Hwy 133, under logs, 16 May 72, JWR, 1-1-4. *Hwy 500, .81 km n of Denbigh, under logs, 16 May 72, JWR, 6-1-4. Hwy 7, Kaladar, e.e., under lumber in school yard, 16 May 72, JWR, 0-0-2. MANITOULIN DIST. *Hwy 68, 3.22 km n of Little Current, under logs and lumber, 13 May 72, JWR & JEM, 0-0-8. MIDDLESEX CO. Judd (1964). *Hwy 73, .97 km n of Mossley, under logs, 4 May 72, JWR, 20-2-8. *Hwy 7, 2.26 km s of Parkhill, under fence posts, 4 May 72, JWR, 0-0-3. MUSKOKA DIST. *Hwy 11, 1.13 km n of Severn Bridge, under logs, 9 May 72, JWR, 2-0-2. *Hwy 11, 11.45 km n of Severn Bridge, under pine logs, 9 May 72, JWR, 1-1-4. *Hwy 11, 12.9 km s of Bracebridge, under rocks, 9 May 72, JWR, 3-1-2. *Hwy 11, 9.19 km s of Hwy 516, under logs and rocks in wet ditch, 9 May 72, JWR, 3-1-5. *Hwy 11, Melissa, s.e., under paper in ditch, 9 May 72, JWR, 3-1-1. NIAGARA CO. Niagara, 1873, GE, 0-2-0, USNM-1156. Niagara, 1873, GE, 0-1-3, USNM-1157. *Queen Elizabeth Hwy, 1.13 km w of Ontario St, Grimsby, under log, 1 May 72, JWR, 0-0-1. NIPISSING DIST. *Hwy 17, 5.48 km e of Warren, under logs and dung in pasture, 13 May 72, JWR & JEM, 6-1-7. *Hwy 17, 1.29 km e of Verner, under paper in wet ditch, 13 May 72, JWR & JEM, 3-1-2. Hwy 17, 7.90 km w of North Bay, digging under rocks, 13 May 72, JWR & JEM, 12-4-17. Temagami, Bearls (Hudson Bay Co. Post), 30 Jun 37, JO & GMN, 0-0-1, ROM-I27. NORFOLK CO. Hwy 6, 4.84 km s of Jarvis, under log, 1 May 72, JWR, 0-0-1. *Hwy 24, 3.06 km n of Hwy 6, under logs, 1 May 72, JWR, 5-4-0. *Hwy 3,

11.29 km e of Delhi, under logs, 1 May 72, JWR, 8-9-0. *Hwy Jct 3 and 59, under junk under pine trees, 1 May 72, JWR, 7-1-2. NORTHUMBERLAND CO. *Hwy 45, 10.69 km s of Norwood, under logs around barn, 28 Apr 72, JWR, 14-2-0. Hwy 45, 4.35 km n of Roseneath, under logs, 28 Apr 72, JWR, 4-0-6. *Hwy 45, 9.03 km n of Baltimore, under logs, 28 Apr 72, JWR, 0-1-1. *Hwy 45, 4.84 km n of Baltimore, under logs, 28 Apr 72, JWR, 0-0-2. *Hwy 401, 8.06 km w of Grafton, under log, 15 May 72, JWR, 1-1-2. *Hwy 401, 5.64 km e of Grafton, under logs, 15 May 72, JWR, 4-0-18. *Northumberland-Durham Rd 1, .65 km w of Hwy 33, under log, 15 May 72, JWR, 1-0-0. Lakeport, on pebbly beach of Lake Ontario, 25 Jul 74, DRB, 0-0-1, UW-0001. ONTARIO CO. *Hwy 7, 1.61 km e of Green River, under logs, 26 Apr 72, JWR, 0-0-4. *Hwy 7-12, .81 km n of Myrtle, under logs, 26 Apr 72, JWR, 4-3-0. *Hwy 47, 8.22 km e of Uxbridge, under log in wet area, 9 May 72, JWR, 3-1-0. *Hwy 47, 10.48 km w of Uxbridge, digging, 9 May 72, JWR, 10-4-1-1. *Hwy 7-12, 3.22 km n of Blackwater, under logs, 9 May 72, JWR, 2-3-1. Dufferin Creek area, day after use of Lampricide, 11 May 71, TY, 0-0-1, ROM-I41. OXFORD CO. *Hwy 59, 6.54 km n of Hwy 3, under logs, 1 May 72, JWR, 6-0-2. *Hwy 59, 1.13 km s of Curries, under junk, 1 May 72, JWR, 0-0-4. *Hwy 59, 1.77 km e of Burgesville, under logs and wood chips, 1 May 72, JWR, 2-2-4. *Hwy 97, Washington, w.e., under log, 3 May 72, JWR, 0-0-1. *Hwy 59, .32 km n of Hickson, under logs, 3 May 72, JWR, 0-4-2. PARRY SOUND DIST. *Hwy 11, 4.03 km s of Scotia Rd, under logs, 9 May 72, JWR, 0-3-1. Hwy 11, 3.87 km s of Katrine, under log, 9 May 72, JWR, 0-0-0-1. *Hwy 11, 4.52 km s of Burks Falls, under logs, 9 May 72, JWR, 7-11-0. Hwy 520, 5 km w of Burks Falls, under log, 9 May 72, JWR, 0-1-0. *Hwy 520, Magnetawan, e.e., under rocks, 9 May 72, JWR, 3-0-3. *Hwy 124, 3.55 km w of Hwy 69, under logs, 9 May 72, JWR, 0-0-5. PEEL CO. *Hwy 5, .81 km e of Dixie Rd, under logs, 29 Apr 72, JWR, 5-0-1. *Hwy 5, 4.35 km w of Hwy 10, under mattress, 29 Apr 72, JWR, 0-0-6. *Hwy 24, 8.87 km n of Erin, under log and dung, 29 Apr 72, JWR, 0-4-0. *Hwy 9, 4.35 km e of Mono Mills, under logs, 29 Apr 72, JWR, 0-1-5. Hwy 401, 1.3 km w of Dixie Rd, wet ditch, 4 May 73, JWR, 0-0-3. PERTH CO. .81 km n of Fullarton, under rock, 29 Apr 72, DWR & LWR, 0-0-1. *Mitchell, next to Collegiate, under logs, 4 May 72, JWR & DWR, 8-0-6. PETERBOROUGH CO. *Hwy 28, Lakefield College, waterfront under logs, 27 Apr 72, JWR & CWR, 4-0-1. *Hwy 504, 1.61 km e of Apsley, under logs, 27 Apr 72, JWR, 1-1-1. *Hwy 504, .81 km w of Lasswade, old house site under logs, 27 Apr 72, JWR, 10-0-7. PRINCE EDWARD CO. *Hwy 33, 1.61 km s of Carrying Place, under log in wet ditch, 15 May 72, JWR, 11-3-11. *Hwy 33, Consecon, n.e., dump, 15 May 72, JWR, 4-0-1. Hwy 33, Bloomfield, e.e., under paper, 15 May 72, JWR, 1-0-1. *Hwy 49, Picton, n.e., under leaf pile, 15 May 72, JWR, 4-0-1. RENFREW CO. *Hwy 500, 12.26 km n of Denbigh, under logs, 16 May 72, JWR, 3-1-6. RUSSELL CO. *2.42 km s of Limonges, under lumber at old house site, 11 May 72, JWR, 0-0-2. SIMCOE CO. *Hwy 89, 5.48 km e of Alliston, under logs, 2 May 72, JWR, 12-0-5. *Hwy 89, 1.94 km e of Rosemont, under logs in wet ditch, 2 May 72, JWR, 0-0-2. Barrie, 14 Greenfield, under logs and digging in garden, 7 May 72, JWR & MKG, 35-3-4. *Barrie, park opposite 14 Greenfield, under rocks and in grass clippings, 7 May 72, JWR & GWA, 0-1-9. *Hwy 400, .81 km s of Hwy 103, under logs, 9 May 72, JWR, 2-1-1. SUDBURY DIST. *Hwy 17, 3.06 km e of Nairne Centre, under logs in ditch, 13 May 72, JWR & JEM, 0-0-5. VICTORIA CO. *Hwy 7, 7.26 km e of Hwy 35, under logs, 26 Apr 72, JWR, 5-3-3. *Hwy 7, 1.94 km w of Hwy 46, under logs and dung, 9 May 72, JWR, 8-6-13. *Hwy 46, 3.71 km n of Hwy 7, under logs, 9 May 72, JWR, 1-3-1. *Hwy 46, .81 km n of Argyle, under logs, 9 May 72, JWR, 15-1-2. *Hwy 46, 8.22 km n of Argyle, under logs, 9 May 72, JWR, 3-6-9. WATERLOO CO. *Hwy 24A, 11.45 km n of Paris, under log, 3 May 72, JWR, 2-1-5. *Hwy 97, Galt, .81 km w of Hwy 24A, under grass clippings, 3 May 72, JWR, 1-0-3. *Hwy 97, 3.71 km w of Hwy 401, under logs, 3 May 72, JWR, 4-1-3. Hwy 85, 1.45 km n of St. Jacobs, digging, 3 May 72, JWR, 0-1-4. Waterloo, Laurel Creek Conservation Area, 3 Aug 74, DPS, 2-1-6, UW-0007. Waterloo, Amos Ave., 1 Sep 75, DPS, 0-0-0-2, UW-0003. WELLINGTON CO. *Hwy 6, 2.26 km n of Aberfoyle, under logs, 29 Apr 72, JWR, 3-1-4. *Hwy 24, 4.03 km n of Erin, under logs, 29 Apr 72, JWR, 4-2-1. *Hwy 24, 5.16 km e of Eramosa, under logs in cedar (*Thuja occidentalis*) woodlot, 29 Apr 72, JWR, 11-5-2. Hwy 6, University of Guelph, on wet driveway behind the Soil Science Building, 2 May 72, JWR, 3-0-11. *Hwy 6, 1.13 km s of Ennotville, under paper in wet ditch, 2 May 72, JWR, 6-0-1. WENTWORTH CO. *Hwy 6, 5.32 km n of Hwy 5, under log, 29 Apr 72, JWR, 0-0-1. YORK CO. Hwy 27, 8.55 km s of Hwy 9, under logs, 29 Apr 72, JWR, 8-0-1. *Edenbrook Park, Islington, under rocks and logs near stream bank, 30 Apr 72, JWR & DWR, 9-5-1. Hwy 27, Bell's Lake, under lumber, 2 May 72, JWR, 5-0-4. Edenbrook Park, Islington, quantitative study 1, formalin, 18 May 72, JWR, 10-0-3. Edenbrook Park, Islington, quantitative study 2, formalin, 18 May 72, JWR, 2-1-6. Toronto, GE, 0-0-4, USNM-4563. Toronto, GE, 0-4-0, USNM-4564.

Aporrectodea turgida (Eisen, 1873)
Pasture worm   Ver du pâturage
(Fig. 14)

1873 *Allolobophora turgida* Eisen, Öfv. Vet.-Akad. Förh. Stockholm 30(8): 46.
1874 *Allolobophora turgida* (part.)–Eisen, Öfv. Vet.-Akad. Förh. Stockholm 31(2): 43.
1946 *Allolobophora caliginosa*–Evans, Ann. Mag. Nat. Hist., ser. 11, 13: 100, 101.
1947 *Allolobophora caliginosa*–Černosvitov and Evans, Linn. Soc. Lond., Syn. British Fauna, no. 6: 13.
1952 *Allolobophora caliginosa* + *A. molita* Gates, Breviora, no. 9: 1, 3.
1959 *Allolobophora caliginosa*–Zicsi, Acta Zool. Hung. 5(1–2): 172.
1964 *Allolobophora caliginosa* (in toto)–Gerard, Linn. Soc. Lond., Syn. British Fauna, no. 6: 27.
1969 *Allolobophora caliginosa* (part.)–Støp-Bowitz, Nytt. Mag. Zool. 17(2): 191.
1970 *Allolobophora caliginosa*–Zajonc, Biol. Práce 16(8): 23.
1972 *Allolobophora caliginosa* f. *typica* (part.) + *A.c.*f. *trapezoides* (part.)–Edwards and Lofty, Biol. earthworms, p. 217.
1972 *Nicodrilus caliginosus caliginosus*–Bouché, Inst. Natn. Rech. Agron., p. 326.
1972 *Allolobophora turgida*–Gates, Bull. Tall Timbers Res. Stn. 12: 89.
1973 *Allolobophora caliginosa* f. *typica* (part.)–Plisko, Fauna Polski, no. 1, p. 107.
1975 *Aporrectodea turgida*–Reynolds, Megadrilogia 2(3): 3.

Diagnosis
Length 60–85 mm, diameter 3.5–5.0 mm, segment number 130–168, prostomium epilobic, first dorsal pore 12/13 or 13/14. Clitellum xxvii, xxviii, xxix–xxxiv, xxxv. Tubercula pubertatis xxxi–xxxiii. Setae closely paired, $AA:AB:BC:CD:DD = 3:1:2:\frac{2}{3}:10$. Genital tumescences contain *a* and *b* only in xxx, xxxii–xxxiv, and frequently in xxvii. Male pores on xv between *b* and *c*. Seminal vesicles, four pairs in 9–12. Spermathecae, two pairs, with short ducts opening at level *cd* in 9/10 and 10/11. Colour, unpigmented with the region anterior to the crop flesh pink and the remaining segments pale grey, or occasionally with light pigmentation on the dorsal surface. Body cylindrical.

Discussion
Aporrectodea turgida, along with the previous two species, has given oligochaetologists great difficulty until a recent publication by Gates (1972a). This species was first recorded from Ontario (Niagara County) by Eisen in 1874. Four specimens labelled *Ap. turgida* from Niagara were sent by Eisen to the United States National Museum (cat. no. 1157). One specimen may be an aberrant individual of *Ap. turgida*, but the other three appear to be *Ap. tuberculata* (Gates, 1972a). I have subsequently examined these specimens and confirm this report.

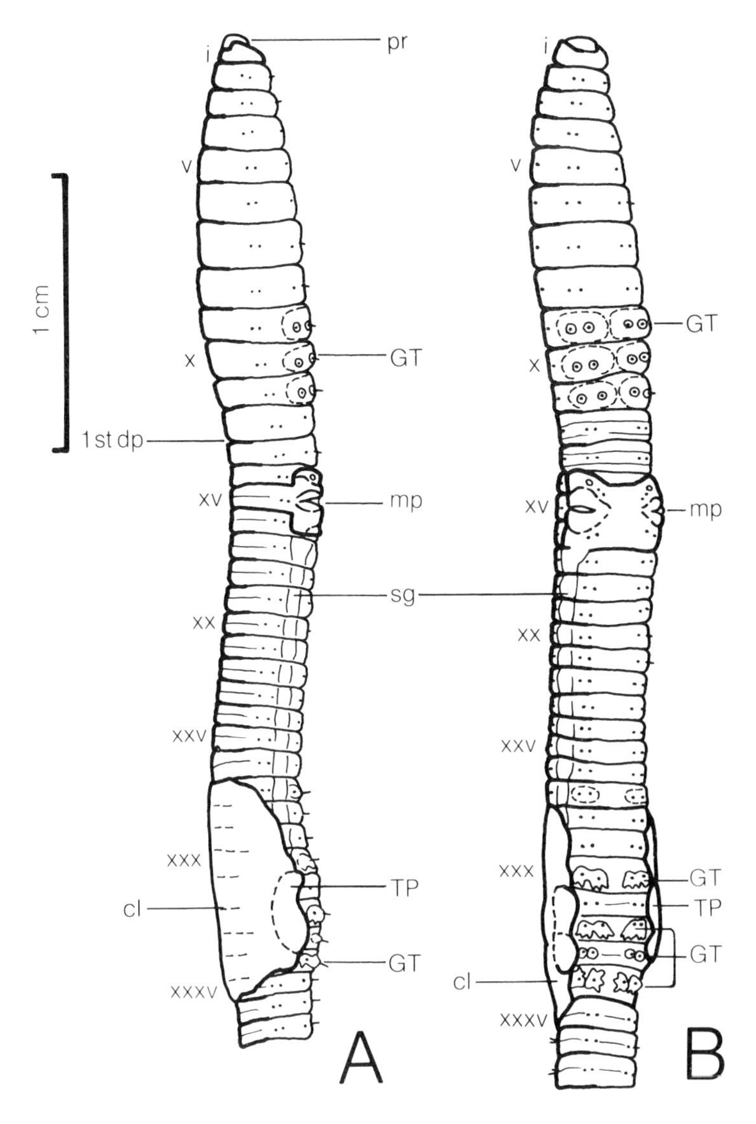

Fig. 14 External longitudinal views of *Aporrectodea turgida* showing taxonomic characters. A. Lateral view. (ONT: Grey Co., cat. no. 7988) B. Ventral view. (ONT: Parry Sound Dist., cat. no., 7891)

Biology

Gates (1972c) records this species from a variety of habitats, including gardens, fields, turf, forest humus, compost, banks of springs and streams, wasteland and city dumps, and from a cave in West Virginia. The widespread distribution and variety of utilized habitats for *Ap. turgida* were recorded also by Eaton (1942), Černosvitov and Evans (1947), Murchie (1956), and Gerard (1964). The amazing water tolerance of this species was reported by Guild (1951) and Reynolds et al. (1974), in Scotland and the United States respectively. The variety of habitats utilized by *Ap. turgida* in Ontario can be seen from the collection data.

Activity may be year round but in most North American areas aestivation and hibernation are probably climatically imposed (Gates, 1972c). Feeding and cast deposition occur below the surface. Reproduction is obligatorily amphimictic with copulation beneath the soil surface (Reynolds, 1974c).

Ap. turgida is potentially a useful species for fish bait where available in sufficient numbers.

Range

Known from Europe, North America, South America, Asia, South Africa, and Australia (Gates, 1972c). Also from Iceland (Backlund, 1949).

North American Distribution

British Columbia (Gates, 1972a), New Brunswick (Reynolds, 1976d), Newfoundland (Gates, 1972a), Nova Scotia (Reynolds, 1975a, 1976a), Ontario (Eisen, 1874), Prince Edward Island (Reynolds, 1975c), Québec (Reynolds, 1975b, d, e, 1976c), Alaska (Gates, 1972a), Arizona (Gates, 1967), California (Gates, 1967), Colorado (Gates, 1967), Connecticut (Reynolds, 1973c), Delaware (Reynolds, 1973a), Florida (Gates, 1972c), Georgia (Gates, 1972a), Idaho (Gates, 1967), Illinois (Garman, 1888), Indiana (Gates, 1972a), Iowa (Gates, 1972a), Kentucky (Gates, 1972a), Louisiana (Harman, 1952), Maine (Gates, 1972a), Maryland (Reynolds, 1974b), Massachusetts (Reynolds, 1977), Michigan (Gates, 1972a), Minnesota (Gates, 1972a), Montana (Reynolds, 1972c), Nevada (Gates, 1967), New Hampshire (Gates, 1972a), North Carolina (Garman, 1888), Ohio (Gates, 1972a), Oregon (Gates, 1972a), Pennsylvania (Gates, 1972a), Rhode Island (Reynolds, 1973b), Vermont (Gates, 1972a), Virginia (Gates, 1972a), West Virginia (Gates, 1959), Wisconsin (Ude, 1885). New records: Manitoba, Arkansas, Mississippi.

Ontario Distribution (Fig. 15)

ALGOMA DIST. *Hwy 17, .48 km e of Spanish, under log in ditch, 13 May 72, JWR and JEM, 0-0-1. BRANT CO. *Hwy 54, .64 km w of Onondaga, under logs in wet area, 3 May 72, JWR, 13-4-3. BRUCE CO. *Hwy 4, 6.77 km s of Teeswater, under logs, 5 May 72, JWR, 4-1-9. *Hwy 4, .48 km s of Hwy 9, under logs, 5 May 72, JWR, 2-2-14. Hwy 9, .81 km w of Bervie, under logs in pasture, 5 May 72, JWR, 0-0-4. *Hwy 21, 6.94 km n of Kincardine, under logs, 5 May 72, JWR, 7-1-6. Hwy 21, 2.58 km n of Tiverton, under lumber, 5 May 72, JWR, 16-2-1. *Hwy 21, 1.61 km e of Elsinore, under log, 5 May 72, JWR, 0-0-1. CARLETON CO. *Ottawa-Carleton Rd 35, 2.42 km s of Leonard, under logs, 11 May 72, JWR, 16-8-10. COCHRANE DIST. Reynolds, (1972a). DUFFERIN CO. *Hwy 89, 10 km e of Primrose, wet ditch, 2 May 72, JWR, 7-0-9. *Hwy 9, 10.16 km w of Orangeville, under junk, 2 May 72, JWR, 3-1-6. *Hwy 10-24, 8.06 km n of Orangeville, under logs, 2 May 72, JWR, 3-2-0. DUNDAS CO. *Hwy 2, 5 km w of Iroquois, under log, 11 May 72, JWR, 7-0-1. *Hwy 2, 5.48 km e of Iroquois, under log, 11 May 72, JWR, 0-0-1. *Hwy 31, 3.55 km n of Morrisburg, under lumber, 11 May 72, JWR, 4-1-10. *Hwy 43, 1.77 km e of Hallville, under logs, 11 May 72, JWR, 2-2-7. DURHAM CO. *Hwy 401, .16 km w of Liberty Rd, Bowmanville, digging under log next to railroad track, 15 May 72, JWR, 0-1-4-1. *Hwy 401, .32 km e of Mill St, Newcastle, under logs, 15 May 72, JWR, 1-1-6. ELGIN CO. *Hwy 73, 3.06 km s of Harrietsville, under rotten log, 4 May 72, JWR, 10-4-2. *Hwy 73, 6.77 km n of Aylmer, under logs, 4 May 72, JWR, 5-5-14. Hwy 3, 5.16 km w of Talbotville, under log, 4 May 72, JWR, 2-0-1. Hwy 3, 7.9 km w of Frome, under logs, 4 May 72, JWR, 3-2-5. ESSEX CO. Hwy 3, 4.52 km w of Wheatley, under logs, 4 May 72, JWR, 1-1-2. *Hwy 2, 8.55 km e of St. Joachim, ditch, 4 May 72, JWR, 0-1-8. *Belle River, s.e., under logs in dump, 4 May 72, JWR, 7-2-7. FRONTENAC CO. *Hwy 2, .97 km w of Westbrook, under logs, 16 May 72, JWR, 3-0-9. GLENGARRY CO. *Hwy 401, 7.42 km w of Summerstown Rd, under logs, 11 May 72, JWR, 13-2-13. Hwy 34, 2.58 km n of Lancaster, under dung, 11 May 72, JWR, 1-1-0. *Hwy 34, 5.64 km n of Alexandria, under logs, 11 May 72, JWR, 1-1-3. GRENVILLE CO. *Hwy 2,

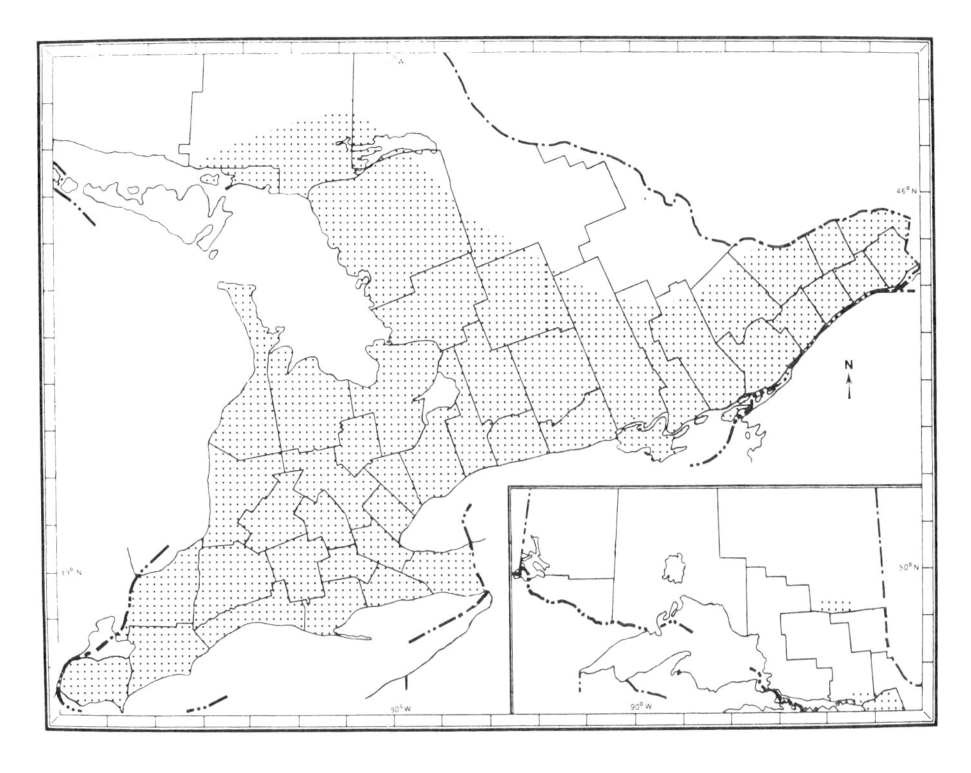

Fig. 15 The known Ontario distribution of *Aporrectodea turgida.*

Johnstown, w.e., under logs, 11 May 72, JWR, 2-0-3. Hwy 2, Cardinal, w.e., under log, 11 May 72, JWR, 0-0-1. *Hwy 43, 3.87 km e of Kemptville, under logs, 11 May 72, JWR, 0-0-3. GREY CO. *Hwy 21, 7.26 km w of Springmount, under rocks, 5 May 72, JWR, 10-1-4. *Hwy 6, 6.94 km s of Dornoch, under logs, 5 May 72, JWR, 7-0-4. *Hwy 4, 8.22 km e of Durham, under logs, 5 May 72, JWR, 4-2-5. HALDIMAND CO. *Hwy 3, 9.03 km e of Dunnville, wet ditch, 1 May 72, JWR, 6-0-4. *Hwy 3, 6.29 km w of Dunnville, under logs, 1 May 72, JWR, 0-0-2. *Hwy 3, 6.13 km w of Cayuga, under logs, 1 May 72, JWR, 3-1-4. *Hwy 3, 2.74 km w of Nelles Corners, under logs, 1 May 72, JWR, 7-2-8. *Hwy 54, 2.26 km n of Caledonia, digging, 3 May 72, JWR, 2-2-4. HALIBURTON CO. *Reynolds (1972a). HALTON CO. Hwy 5, 5.81 km e of Hwy 25, under paper and leaves in ditch, 29 Apr 72, JWR, 2-4-5. *Hwy 7, 4.84 km e of Georgetown, under burnt logs in soil, 4 May 73, JWR, 4-0-5. *Hwy 401, 4.51 km e of Trafalger Rd, under logs, 4 May 73, JWR, 1-2-1. HASTINGS CO. Reynolds (1972a). *Hwy 62, 1.61 km n of Hwy 7, under logs in wet ditch, 27 Apr 72, JWR, 2-2-3. HURON CO. *Hwy 83, 6.77 km w of Russeldale, under logs, 5 May 72, JWR, 15-2-0. *Hwy 4, 1.61 km n of Exeter, wet ditch, 5 May 72, JWR, 8-0-2. Hwy 4, 1.29 km n of Hensall, under corn (*Zea maize*) cobs in field, 5 May 72, JWR, 1-1-4. Hwy 4, 5.16 km s of Clinton, under logs, 5 May 72, JWR, 5-1-5. *Hwy 4, 2.74 km n of Clinton, under logs, 5 May 72, JWR, 2-3-8-1. KENT CO. *River Rd, 6.54 km w of Chatham, under logs, 23 Apr 72, JWR & TW, 1-2-1. Hwy 3, 1.27 km w of Palmyra, under logs, 4 May 72, JWR, 9-9-11. LAMBTON CO. *Hwy 21, 6.45 km n of Dresden, under logs, 4 May 72, JWR, 0-0-3. Hwy 21, 11.29 km n of Dresden, under dung in pasture, 4 May 72, JWR, 0-1-5. Hwy 21, .64 km s of Petrolia, ditch, 4 May 72, 6-1-5. *Hwy 21, Edy's Mills, s.e., under rail road ties, 4 May 72, JWR, 0-0-2. Hwy 21, 2.1 km n of Wyoming, under logs in wet ditch, 4 May 72, JWR, 1-2-7. LANARK CO. *Hwy 43, 12.1 km e of Smiths Falls, under logs, 11 May 72, JWR, 1-0-1. *Hwy 7, 5.32 km e of Maberly, under dung in pasture, 11 May 72, JWR, 0-0-1. LEEDS CO.

*Hwy 15, 2.26 km n of Seeley's Bay, under logs and concrete blocks in ditch, 16 May 72, JWR, 7-0-17. *Hwy 15, 5.64 km n of Morton, wet ditch, 16 May 72, JWR, 3-1-1. *Hwy 15, 4.84 km n of Elgin, digging, 16 May 72, JWR, 6-1-2. *Hwy 15, .97 km s of Portland, under logs, 16 May 72, JWR, 22-0-17. *Hwy 15, 4.19 km s of Lombardy, under logs and crawling on the surface during rain, 16 May 72, JWR, 2-2-3. *Hwy 15, 7.1 km n of Lombardy, under paper in ditch, 16 May 72, JWR, 0-1-2. MANITOULIN DIST. *Hwy 68, 5.97 km n of Birch Island, under rock, 13 May 72, JWR & JEM, 0-0-1. *Hwy 68, Birch Island, s.e., under logs, 13 May 72, JWR & JEM, 7-7-10. *Hwy 68, 2.42 km s of Birch Island, under logs, 13 May 72, JWR & JEM, 3-0-6. Hwy 68, 3.22 km n of Little Current, under logs and lumber, 13 May 72, JWR & JEM, 1-0-4. MIDDLESEX CO. Judd (1964). *Hwy 7, 2.26 km s of Parkhill, under fence posts, 4 May 72, JWR, 3-0-13. MUSKOKA DIST. Hwy 11, Huntsville, 620 m s of Ravenscliffe Rd, under logs, 9 May 72, JWR, 4-1-1. NIAGARA CO. Eisen (1874). *Queen Elizabeth Hwy, 1.13 km w of Ontario St, Grimsby, under logs and lumber, 1 May 72, JWR, 3-1-2. *Niagara Co. Rd 12, 3.06 km s of Grimsby, under paper in wet ditch, 1 May 72, JWR, 0-0-2. *Hwy 20, 1.61 km e of Smithville, under logs at old house site, 1 May 72, JWR, 5-6-5. NIPISSING DIST. *Hwy 17, 8.39 km e of Warren, digging at house site, 13 May 72, JWR & JEM, 9-2-9. NORFOLK CO. *Hwy 6, 4.84 km s of Jarvis, under logs, 1 May 72, JWR, 3-2-1. *Hwy 24, 3.06 km n of Hwy 6, under logs, 1 May 72, JWR, 0-3-0. NORTHUMBERLAND CO. *Hwy 45, 4.84 km n of Baltimore, under logs, 28 Apr 72, JWR, 6-1-7. ONTARIO CO. *Hwy 7, 1.61 km e of Green River, under logs, 26 Apr 72, JWR, 13-3-9. *Hwy 47, 10.48 km w of Uxbridge, digging, 9 May 72, JWR, 0-0-1. OXFORD CO. *Hwy 59, 1.94 km s of Norwich, under log, 1 May 72, JWR, 6-3-0. *Hwy 59, 1-1-3 km s of Curries, under junk, 1 May 72, JWR, 4-2-7. *Hwy 59, 1.77 km e of Burgessville, under logs and wood chips, 1 May 72, JWR, 0-5-9. *Hwy 97, Washington, w.e., under logs, 3 May 72, JWR, 3-1-4. *Hwy 59, .32 km n of Hickson, under logs, 3 May 72, JWR, 13-6-6. PARRY SOUND DIST. *Hwy 520, 10.32 km e of Magnetawan, under lumber, 9 May 72, JWR, 2-0-6. *Hwy 124, 2.26 km w of Hwy 520, under log, 9 May 72, JWR, 0-0-1. PEEL CO. *Hwy 5, 4.35 km w of Hwy 10, under mattress, 29 Apr 72, JWR, 3-1-1. PERTH CO. .81 km n of Fullarton, under rocks, 29 Apr 72, DWR & LWR, 6-0-5. Hwy Jct 7, 8 and 59, Shakespeare, under logs, 3 May 72, JWR, 0-0-5. *Mitchell, next to Collegiate, under logs, 4 May 72, JWR & DWR, 6-4-5. PETERBOROUGH CO. *Hwy 28, 5.48 km n of Burleigh Falls, under logs, 27 Apr 72, JWR, 3-0-1. PRESCOTT CO. *Hwy 34, 5.48 km s of Vanleek Hill, under logs, 11 May 72, JWR, 7-1-4. *Hwy 34, 5.81 km n of Vanleek Hill, under logs, 11 May 72, JWR, 13-0-15. Hwy 17, .64 km e of Plantagenet, under log, 11 May 72, JWR, 0-0-1. PRINCE EDWARD CO. *Hwy 33, Consecon, n.e., dump, 15 May 72, JWR, 0-0-1. RUSSELL CO. *Hwy 17, 7.74 km e of Rockland, under paper in wet ditch, 11 May 72, JWR, 0-0-1-1. *Hwy 17, 4.35 km w of Rockland, under logs, 11 May 72, JWR, 4-0-1. *2.9 km e of Embrun, under logs, 11 May 72, JWR, 2-4-5. SIMCOE CO. *Hwy 89, 1.94 km e of Rosemont, under logs in wet ditch, 2 May 72, JWR, 7-10-14. Hwy 27, 3.22 km s of Newton Robinson, wet ditch, 2 May 72, JWR, 2-3-2. *Barrie, park opposite 14 Greenfield, under rocks and grass clippings, 7 May 72, JWR & GWA, 0-0-10. *Hwy 11, 15.16 km s of Severn Bridge, under logs, 9 May 72, JWR, 4-0-4. *Hwy 400, .81 km s of Hwy 103, under logs, 9 May 72, JWR, 3-1-1. STORMONT CO. *Hwy 43, 1.29 km w of Finch, under logs, 11 May 72, JWR, 4-0-16. Hwy 43, 3.71 km w of Monkland, under logs, 11 May 72, JWR, 0-1-8. *Hwy 138, 3.06 km n of St. Andrews West, under log, 11 May 72, JWR, 8-2-4. SUDBURY DIST. *Hwy 17, 3.06 km e of Nairne Centre, under logs in ditch, 13 May 72, JWR & JEM, 11-1-4. Hwy 68, Espanola, n.e., under log and rock, JWR & JEM, 1-0-1. VICTORIA CO. *Hwy 35, 8.71 km s of Hwy 7, under log, 26 Apr 72, JWR, 3-0-4. *Hwy 7, 1.94 km w of Hwy 46, under log and dung, 9 May 72, JWR, 0-2-1. *Hwy 46, .81 km n of Argyle, under logs, 9 May 72, JWR, 18-1-2. Hwy 48, Bolsover, e.e., under paper in wet ditch, 9 May 72, JWR, 0-0-1. WATERLOO CO. *Hwy 97, Galt, .81 km w of Hwy 24A, under grass clippings, 3 May 72, JWR, 5-0-2. Hwy 85, 1.45 km n of St. Jacobs, digging, 3 May 72, JWR, 3-1-1. Hwy 86, West Montrose, under logs, 3 May 73, JWR, 6-2-3. Waterloo, Amos Ave., 1 Sep 75, DPS, 0-9-4, UW-0003. WELLINGTON CO. *Hwy 6, 2.26 km n of Aberfoyle, under logs, 29 Apr 72, JWR, 7-0-2. *Hwy 6, 10.64 km s of Arthur, under logs, 2 May 72, JWR, 19-2-8. WENTWORTH CO. Reynolds (1972a). *Hwy 5, Waterdown, e.e., edge of corn (*Zea maize*) field, 29 Apr 72, JWR, 21-1-4. YORK CO. *Hwy 48, .32 km n of Steeles Avenue, under logs, 26 Apr 72, JWR, 11-2-4. *Hwy 27, 1.45 km n of Hwy 7, under logs, 29 Apr 72, JWR, 2-1-3. *Edenbrook Park, Islington, under logs and rocks near stream bank, 30 Apr 72, JWR & DWR, 54-9-4. Edenbrook Park, Islington, quantitative study 1, formalin, 18 May 72, JWR, 2-0-3. Edenbrook Park, Islington, quantitative study 2, formalin, 18 May 72, JWR, 2-0-2. Kings Township, 20 Sep 41, JRD, 6-0-1, ROM-I25.

Genus *Bimastos* Moore, 1893

1893 *Bimastos* Moore, Zool. Anz. 16: 333.
1895 *Bimastos*–Moore, J. Morph. 10(2): 473.
1900 *Helodrilus* (*Bimastus*)–Michaelsen, Das Tierreich, Oligochaeta 10: 501.
1930 *Bimastus*–Stephenson, Oligochaeta, p. 913.
1969 *Bimastos*–Gates, J. Nat. Hist. Lond. 9: 306.
1972 *Bimastos*–Gates, Trans. Amer. Philos. Soc. 62(7): 86.
1975 *Bimastos*–Gates, Megadrilogica 2(1): 4.

Type Species
Bimastos palustris Moore, 1895.

Diagnosis
Calciferous gland, without marked widening in xi–xii, opening into gut in x through paired vertical sacs. Calciferous lamellae continued onto posterior walls of sacs. Gizzard, mainly in xvii (ending at ?). Extraoesophageal trunks, joining dorsal trunk in xii. Hearts, in vii–xi. Nephridial bladders, *U*-shaped, closed ends laterally, ducts passing into parietes near *B*. Nephropores, inconspicuous, irregularly alternating (and with asymmetry), between levels somewhat above *B* and well above *D*. Setae, closely paired. Dorsal pores, present from region of 5/6. Male pores, equatorial in xv, in atrial chambers invaginated deeply into the coelom and bearing acinous glands. Female pores, equatorial on xiv, shortly above *B*. Holandric, seminal vesicles in xi–xii. Spermathecae, tubercula pubertatis and TP glands, lacking. Prostomium epilobic. Pigment red. (after Gates, 1972c: 86; 1975a: 4).

Discussion
This nearctic genus does not normally occur in Canada. The one collection of four specimens was from an accidental introduction (cf. *B. parvus*). Over the years there has been confusion over the spelling of this and other generic names (cf. *Octolasion*, p. 104).

Bimastos parvus (Eisen, 1874)

American bark worm Ver américain de l'écorce
(Fig. 16)

1874 *Allolobophora parva* Eisen, Öfv. Vet.-Akad. Förh. Stockholm 31(2): 46.
1889 *Lumbricus* (*Allobophora*) *parvus*–L. Vaillant, Hist. Nat. Annel. 3(1): 142.
1893 *Dendrobaena constricta* (part.)–Friend, Naturalist, p. 19.
1896 *Allolobophora parvus* (part.: excl. subsp. *A. udei*)–Ribaucourt, Rev. Suisse Zool. 4: 80.
1897 *Allolobophora constricta* var. *germinata* (part.) Friend, Zoologist, ser. 4, 1: 459.

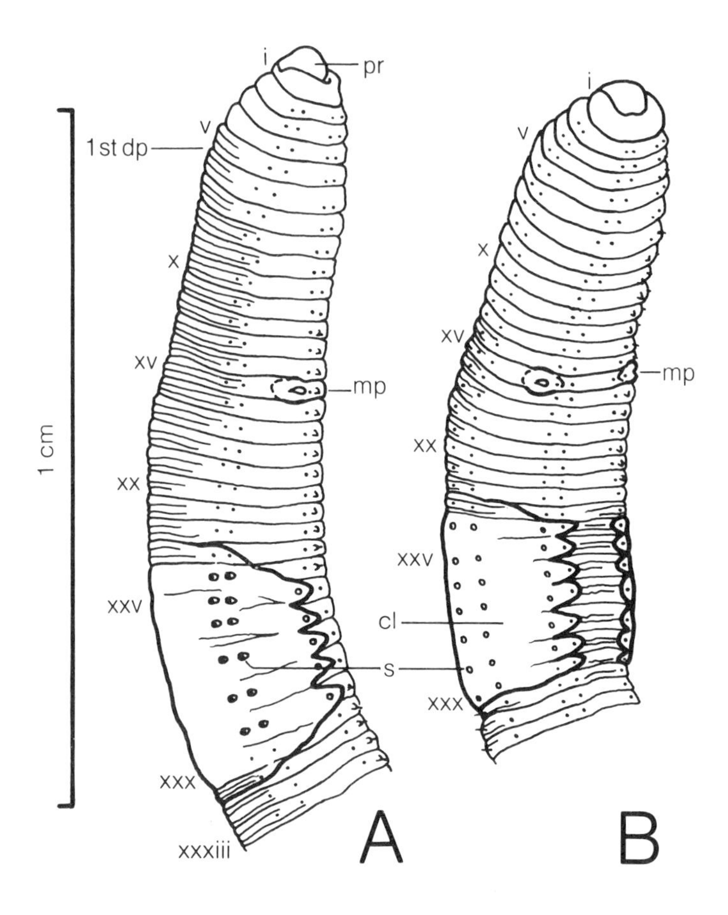

Fig. 16 External longitudinal views of *Bimastos parvus* showing taxonomic characters. A. Lateral view. B. Ventrolateral view. (FL: Franklin Co., cat. no. 3320)

1900 *Allolobophora* (*Bimastus*) *parvus*–Michaelsen, Abh. Nat. Verh. Hamburg 16(1): 10, 14.
1948 *Eisenia parvus* + *Bimastus beddardi*–Pop, Ann. Acad. Sect. Ştiint. Geol. Geogr. Biol., ser. A, 1(9): 123.
1959 *Eisenia parva* (part. ?)–Zicsi, Acta Zool. Hung. 5(1-2): 170.
1970 *Eisenia parva* (part. ?)–Zajonc, Biol. Práce 16(8): 23.
1972 *Eisenia parva* (part. ?)–Bouché, Inst. Natn. Rech. Agron., p. 386.
1972 *Bimastos parvus* (part. ?)–Edwards and Lofty, Biol. earthworms, p. 215.
1972 *Bimastos parvus*–Reynolds, Megadrilogica 1(3): 1.
1972 *Bimastos parvus*–Gates, Trans. Amer. Philos. Soc. 62(7): 87.
1973 *Bimastos parvus* (part. ?)–Plisko, Fauna Polski, no. 1, p. 99.
1974 *Bimastos parvus*–Reynolds, J. Tenn. Acad. Sci. 49(1): 17.

Diagnosis

Length 17–65 mm, diameter 1.5–3.0 mm, segment number 65–97, prostomium epilobic, first dorsal pore 5/6. Clitellum xxiii, xxiv–xxxi, xxxii. Tubercula pubertatis absent or in the form of indefinite ridges on xxiv, xxv, xxvi–xxx. Setae closely paired, $CD = \frac{3}{4}\,AB$, AA slightly greater than BC, $DD = \frac{1}{2}C$. Male pores with small, slightly elevated papillae of a yellowish brown colour on xv. Seminal vesicles in 11 and 12. Spermathecae absent. Colour, reddish dorsally and yellowish ventrally.

Biology

B. parvus, like all the members of the nearctic genus *Bimastos*, is found in close association with decaying logs and leaves, i.e., habitats high in organic matter. This association has been reported by Černosvitov and Evans (1947), Causey (1952), Murchie (1956), and Reynolds et al. (1974). Gates (1972c) recorded it from soils with pH of 7.5 that were wetted with effluent from human habitations, and from gardens, fields, humus, manure, dumps, and litter in caves.

Activity seems to be possible all year under favourable conditions (Gates, 1972c). *B. parvus* is obligatorily, or at least facultatively, parthenogenetic (Reynolds, 1974c). Sperm are rarely present, male sterility is apparent, and spermatophores are unknown (Gates, 1972c).

Range

This species is the only known native North American anthropochore, but is also now known from Europe, North America, South America, Asia, South Africa, and Australia (Gates, 1972c).

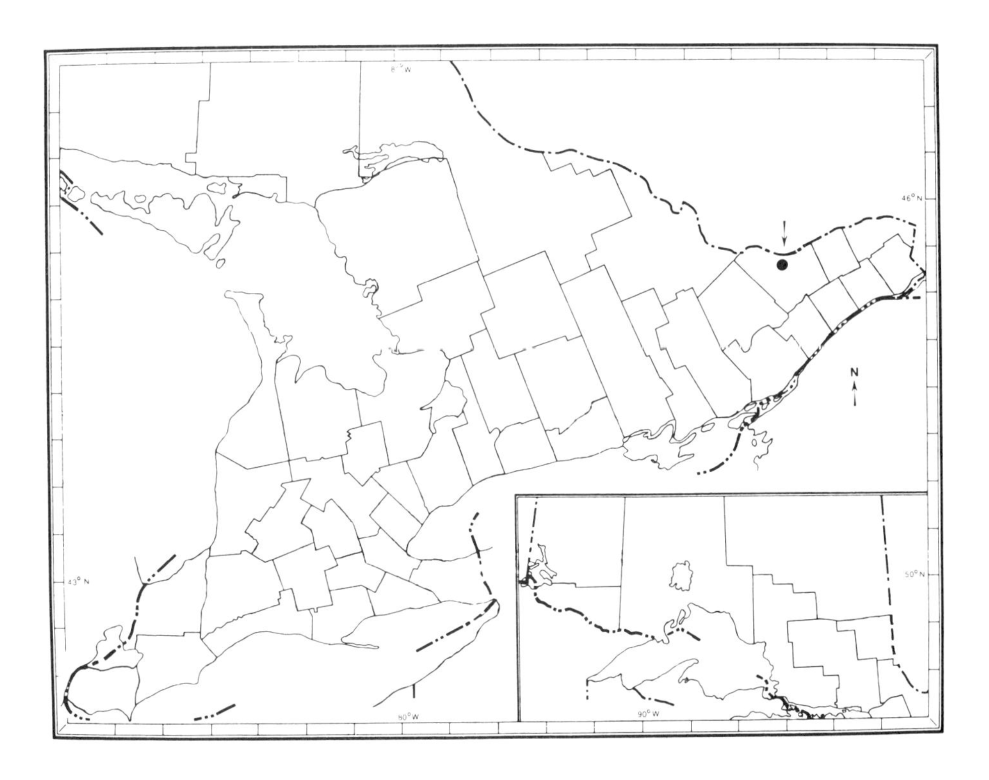

Fig. 17 The known Ontario distribution of *Bimastos parvus*.

North American Distribution

Ontario (Reynolds, 1972a), Arkansas (Causey, 1952), California (Michaelsen, 1900b), Illinois (Gates, 1972c), Kansas (Gates, 1967), Louisiana (Michaelsen, 1900b), Maryland (Reynolds, 1974b), Massachusetts (Reynolds, 1977), Michigan (Smith and Green, 1916), Missouri (Olson, 1936), Montana (Gates, 1972c), Nebraska (Swartz, 1929), Nevada (Gates, 1967), New York (Michaelsen, 1900b), Tennessee (Reynolds, 1974a), Texas (Gates, 1956), Washington (Gates, 1972c), Wyoming (Gates, 1967). New record: Florida.

Ontario Distribution (Fig. 17)

Bimastos parvus has been reported only once from Ontario, by Reynolds (1972a), as an accidental introduction into the Canadian Agricultural Arboretum in Ottawa.

CARLETON CO. Reynolds (1972a).

Genus *Dendrobaena* Eisen, 1873

1873 *Dendrobaena* Eisen, Öfv. Vet.-Akad. Förh. Stockholm 30(8): 53.
1900 *Helodrilus* (*Dendrobaena*) (part.) + *Helodrilus* (*Bimastus*) (part.)–Michaelsen, Das Tierreich, Oligochaeta 10: 488, 501.
1930 *Dendrobaena* (part.)–Stephenson, Oligochaeta, p. 912.
1972 *Dendrobaena* (part.)–Gates, Trans. Amer. Philos. Soc. 62(7): 88.
1972 *Dendrobaena* (part.)–Bouché, Inst. Natn. Rech. Agron., p. 388.
1975 *Dendrobaena*–Gates, Megadrilogica 2(1): 3.

Type Species

Dendrobaena boeckii Eisen by monotypy in 1873 (= *Enterion octaedrum* Savigny, 1826).

Diagnosis

Calciferous gland, without sacs opening into gut at vicinity of 10/11, markedly moniliform in xi–xii. Calciferous sacs, lacking. Gizzard, mainly in xvii. Extraoesophageal trunks, passing to dorsal trunk in vicinity of 9/10. Hearts, in vii–ix. Nephridial bladders, ocarina-shaped, with bluntly rounded end laterally and pointed end mesially, ventral side funnel-shaped and narrowing to pass into parieties at *B*. Nephropores, obvious, behind first few segments in one rank on each side, just above *B*. Setae, not closely paired. Prostomium epilobic. Longitudinal musculature, pinnate. Pigment, red. (after Gates, 1972c: 88 and 1975a: 3).

Discussion

The above diagnosis, after Gates (1972c and 1975a), is for a genus with *D. octaedra* as the type species. But another of Savigny's species, *Enterion rubidum*, has been congeneric with *D. octaedra* for decades almost solely because of similarities in their genitalia. As a result of recent studies based on more conservative somatic anatomy, Omodeo's subgenus *Dendrodrilus* (1956) has been elevated to full genus status with *Enterion rubidum* as the type species (Gates, 1975a: 4).

***Dendrobaena octaedra* (Savigny, 1826)**

Octagonal-tail worm Ver à queue octogonale
(Fig. 18)

1826 *Enterion octaedrum* Savigny, Mém. Acad. Sci. Inst. Fr. 5: 183.
1837 *Lumbricus octaedrus* + *L. vetaedrus* (laps.)–Dugès, Ann. Sci. Nat., ser. 2, 8: 17, 24, 35.
1845 *Lumbricus riparius* (part.) Hoffmeister, Regenwürmer, p. 30.
1849 *Lumbricus flaviventris* R. Leuckart, Arch. Naturg. 15(1): 159.
1871 *Lumbricus puter* (part.) Eisen, Öfv. Vet.-Akad. Förh. Stockholm 27(10): 959.
1873 *Dendrobaena boeckii* Eisen, Öfv. Vet.-Akad. Förh. Stockholm 30(8): 53.
1879 *Lumbricus boeckii*–Tauber, Annul. Danmark, p. 69.
1882 *Dendrobaena camerani* Rosa, Atti Acc. Torino 18: 172.
1885 *Octolasion boeckii*–Örley, Ertek. Term. Magyar Akad. 15(18): 20.
1887 *Allolobophora octaedra*–Rosa, Boll. Mus. Zool. Torino 2(31): 2.
1888 *Dendrobaena octaedra*–Vejdovský, Entwickgesch. Unters., p. 41.
1889 *Lumbricus (Dendrobaena) camerani* + *L. (D.) boeckii* + *L. (D.) octaedrus*–L. Vaillant, Hist. Nat. Annel. 3(1): 113, 118, 119.
1893 *Allolobophora (Dendrobaena) octaedra* (laps.)–Rosa, Mem. Acc. Torino, ser. 2, 43: 424, 437.
1896 *Allolobophora liliputiana* + *A. alpinula* Ribaucourt, Rev. Suisse Zool. 4: 32, 33, 37, 38.
1900 *Helodrilus (Dendrobaena) octaedrus*–Michaelsen, Das Tierreich, Oligochaeta 10: 494.
1948 *Dendrobaena octaedra* f. *typica* + *D. o.* var. *quadrivesiculata* Pop, Ann. Acad. Sect. Ştiint. Geol. Geogr. Biol., ser. A, 1(9): 104, 105, 106.
1964 *Dendrobaena octahedra* (laps.)–Langmaid, Can. J. Soil Sci. 44:34.
1972 *Dendrobaena (Dendrobaena) octaedra*–Bouché, Inst. Natn. Rech. Agron., p. 388.
1974 *Dendrobaena octaedra*–Gates, Bull. Tall Timbers Res. Stn. 15: 16.

Diagnosis

Length 17–60 mm, diameter 3–5 mm, segment number 60–100, prostomium epilobic, first dorsal pore 4/5–6/7. Clitellum xxvii, xxviii, xxix–xxxiii, xxxiv. Tubercula pubertatis usually on xxxi–xxxiii. Setae widely spread, $AA = AB = CD$ and DD is slightly greater. On xvi setae *a* or *b* are found on small genital tumescences. Male pores on xv surrounded by small, often indistinct, glandular papillae. Seminal vesicles in 9, 11, and 12. Spermathecae, three pairs with long ducts on level with setae *d* opening in 9/10–11/12. Body cylindrical, with posterior portion octagonal. Colour, red, dark red to purple.

Biology

According to Gates (1972c), *Dendrobaena octaedra* is found in soils with a pH of 3.0–7.7, mostly in sites little affected by cultivation. Murchie (1956), from his studies in Michigan, characterized three types of habitat for the species: in sod or under moss on stream banks, under logs and leafy debris, or in cool, moist

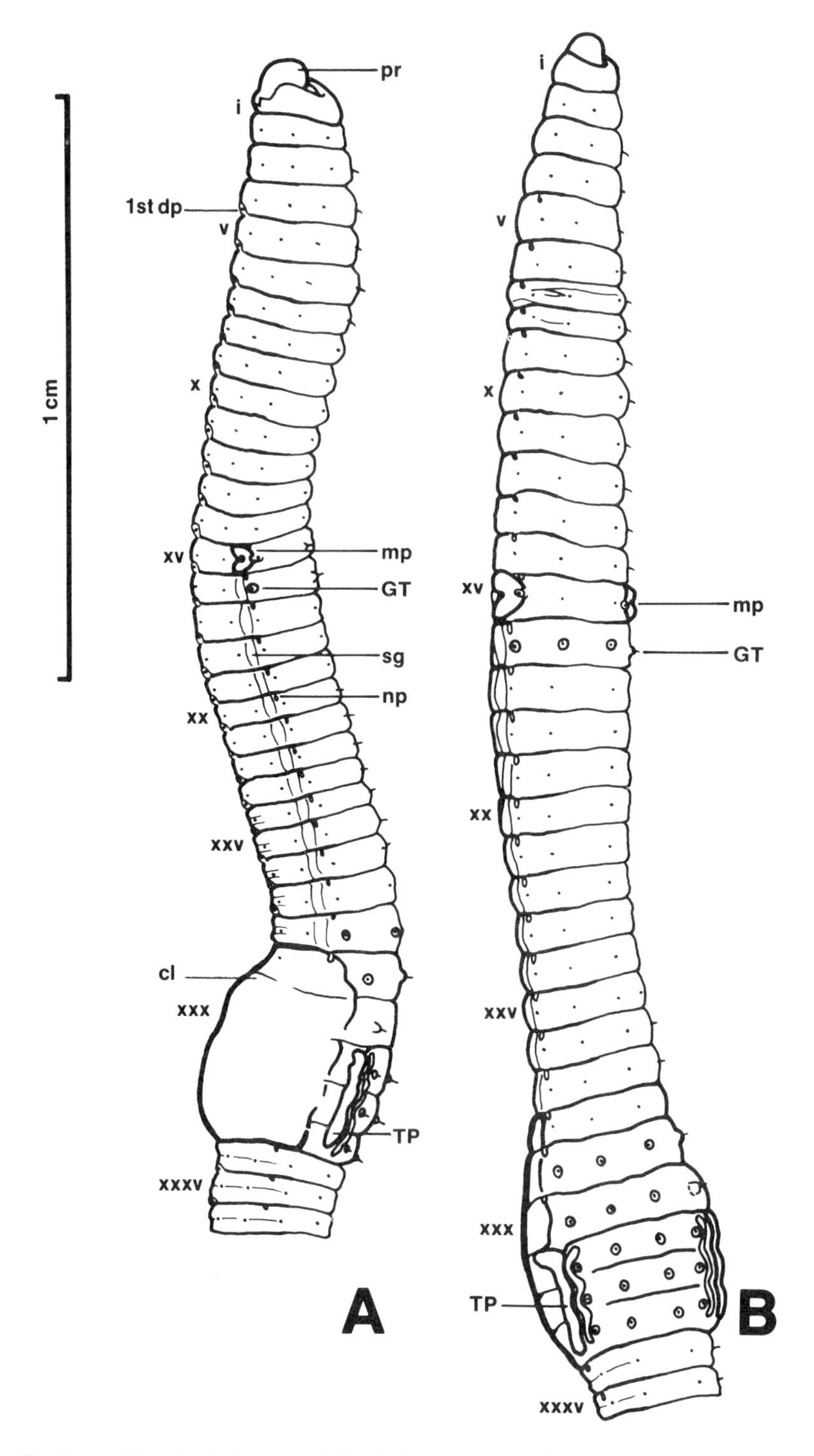

Fig. 18 External longitudinal views of *Dendrobaena octaedra* showing taxonomic characters. A. Lateral view (ONT: York Co., cat. no. 7692) B. Ventral view (ONT: Sudbury Dist., cat. no. 7732)

ravines and upland seepage areas. Gerard (1964) found it often under dung and in soil high in organic matter. These are the types of habitats in which it has been found in Ontario. In Europe *D. octaedra* has been found on mountain tops and in caves, and it is known from botanical gardens and arboretums in Europe and North America.

Under suitable conditions activity, including breeding, is possible the year round. In Maine, and therefore probably in Ontario also, there are two breeding periods. The species is said to be surface living, the upper layers being abandoned only for aestivation and hibernation. Feeding is selective and sand and rock particles are rarely found in the intestine.

Parthenogenetic polymorphism is widespread in *D. octaedra*, probably more so than in any other lumbricid. A detailed discussion of the numerous morphs can be found in a recent paper by Gates (1974b).

In Russia (Kursk) *D. octaedra* has been regarded as an important converter of leaf substances and it is believed mainly responsible for the decomposition of oak leaves (Gates, 1972c). In France, where these worms can be found crawling on bare mountain rocks, the viscous trails are held responsible for trapping lichens and thus initiating a lichen–moss–vascular plant succession (Ribaucourt and Combault, 1906).

Range

A native of Palaearctis, *Dendrobaena octaedra* is now known from Europe, North America, Colombia, Mexico, and Asia. This restriction of a peregrine lumbricid to the Northern Hemisphere is unusual (Gates, 1972c). Also known from Iceland (Backlund, 1949).

North American Distribution

Alberta (Gates, 1974b), British Columbia (Gates, 1974b), New Brunswick (Reynolds, 1976d), Newfoundland (Gates, 1974b), Nova Scotia (Reynolds, 1975a, 1976a), Ontario (Reynolds, 1972a), Prince Edward Island (Reynolds, 1975c), Québec (Reynolds, 1975b, d, e, 1976c), Alaska (Gates, 1974b), Arkansas (Causey, 1952), California (Gates, 1972c), Colorado (Smith, 1917), Connecticut (Reynolds, 1973c), Delaware (Reynolds, 1973a), Georgia (Gates, 1974b), Illinois (Smith, 1928), Indiana (Joyner, 1960), Maine (Gates, 1974b), Maryland (Reynolds, 1974b), Massachusetts (Reynolds, 1977), Michigan (Murchie, 1956), Nebraska (Gates, 1967), New Hampshire (Gates, 1972c), New Jersey (Davies, 1954), New York (Eaton, 1942), North Carolina (Černosvitov, 1942), Ohio (Olson, 1932), Oregon (Gates, 1974b), Pennsylvania (Bhatti, 1965), Rhode Island (Reynolds, 1973b), Tennessee (Reynolds, 1972b), Vermont (Gates, 1949), Virginia (Smith, 1928), Washington (Altman, 1936), West Virginia (Reynolds, 1974b), Wisconsin (Gates, 1972c), Greenland (Levinsen, 1884). New records: Manitoba, Colorado, Minnesota.

Ontario Distribution (Fig. 19)

Dendrobaena octaedra was first reported from Ontario by Reynolds (1972a). This species is found primarily in the Ottawa Valley, extending westward through the Lake Nipissing area.

CARLETON CO. *Ottawa-Carleton Rd 34, Cumberland, under paper, 11 May 72, JWR, 0-0-9. *Ottawa-Carleton Rd 34, .97 km n of Leonard, under paper in wet ditch, 11 May 72, JWR, 1-8-3. HALIBURTON CO. Reynolds (1972a). HALTON CO. *Hwy 401, 4.51 km e of Trafalgar Rd, under logs, 4 May 73, JWR, 0-1-1. KENORA DIST. Rushing River Provincial Park, stream trickle in

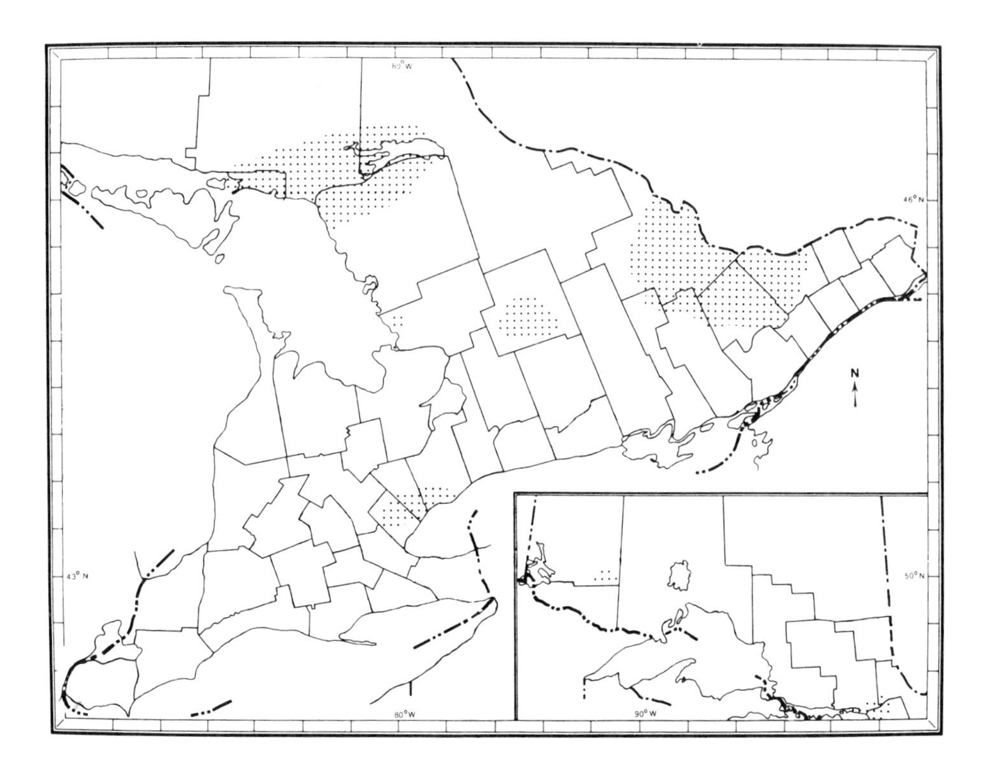

Fig. 19 The known Ontario distribution of *Dendrobaena octaedra*.

campground, 13-14 Jun 71, ROM Field Party, 1-0-0, ROM-135. LANARK CO. *Hwy 7, 5.32 km e of Maberly, under dung in pasture, 11 May 72, JWR, 0-3-1. MUSKOKA DIST. Island in Go Home Lake, under moss in rotting log near pond, 8 Oct 75, RLH, 3-0-0, UW-0001. MANITOULIN DIST. *Hwy 68, 3.06 km n of Birch Island, under logs and paper in ditch, 13 May 72, JWR & JEM, 2-1-1. NIPISSING DIST. Hwy 17, 15.81 km e of Sturgeon Falls, dump, 13 May 72, JWR & JEM, 0-0-1. NORTHUMBERLAND CO. *Hwy 45, Baltimore, under junk, 28 Apr 72, JWR, 1-0-1. PARRY SOUND DIST. *Hwy 124, 4.35 km e of Dunchurch, under logs and rocks, 9 May 72, JWR, 0-3-1. RENFREW CO. Reynolds (1972a). *Hwy 500, 2.42 km s of Hardwood Lake, under logs and dung in pasture, 16 May 72, JWR, 2-1-2. *2.42 km s of Palmer Rapids, under paper in ditch, 16 May 72, JWR, 6-6-15. SUDBURY DIST. *Hwy 17, 3.06 km e of Nairne Centre, under logs in ditch, 13 May 72, JWR & JEM, 2-1-2. YORK CO. Edenbrook Park, Islington, quantitative study 2, formalin, 18 May 72, JWR, 0-0-2.

Genus *Dendrodrilus* Omodeo, 1956

1956 *Dendrobaena (Dendrodrilus)* Omodeo, Arch. Zool. It. 41: 175.
1969 *Dendrobaena* (part.)–Støp-Bowitz, Nytt. Mag. Zool. 17(2): 214.
1972 *Dendrobaena* (part.)–Gates, Trans. Amer. Philos. Soc. 62(7): 88.
1972 *Dendrobaena* (part.)–Bouché, Inst. Natn. Rech. Agron., p. 388.
1973 *Dendrodrilus*–Plisko, Fauna Polski, no. 1, p. 78.
1975 *Dendrodrilus*–Gates, Megadrilogica 2(1): 4.

Type Species
Enterion rubidum Savigny, 1826.

Diagnosis
Calciferous glands, opening into gut ventrally through a pair of sacs posteriorly just in front of insertion of 10/11. Calcierous lamellae continued along lateral walls of sacs. Gizzard, mainly in xvii. Extraoesophageal trunks, passing to dorsal trunk in xii. Hearts, in vii–xi. Nephridial bladders, *U*-shaped loop. Nephropores, inconspicuous, alternating irregularly and with asymmetry on each side between a level above *B* and one above *D*. Setae, not closely paired. Prostomium epilobic. Longitudinal musculature, pinnate. Pigment, red. (after Gates, 1972c: 88 and 1975a: 4).

Discussion
Species of *Dendrodrilus* formerly congeneric with species in *Dendrobaena* because of similarities in genital anatomy are now separated on the basis of differences in their more conservative somatic anatomy.

Dendrodrilus rubidus (Savigny, 1826)
European bark worm   Ver européen de l'écorce
(Fig. 20)

1826 *Enterion rubidum* Savigny, Mém. Acad. Sci. Inst. Fr. 5: 182.
1836 *Lumbricus xanthurus* R. Templeton, Ann. Mag. Nat. Hist. 9: 235.
1837 *Lumbricus rubidus*–Dugès, Ann. Sci. Nat., ser. 2, 8: 17, 23.
1849 ? *Lumbricus valdiviensis* E. Blanchard, Hist. Chile 3: 43.
1867 ? *Hypogeon havaicus* Kinberg, Öfv. Vet.-Akad. Förh. Stockholm 23: 101.
1873 *Allolobophora norvegica* + *A. arborea* + *A. subrubicunda* Eisen, Öfv. Vet.-Akad. Förh. Stockholm 30(8): 48, 49, 51.
1874 *Allolobophora tenuis* Eisen, Öfv. Vet.-Akad. Förh. Stockholm 31(2): 44.
1881 *Allolobophora fraissei* Örley, Zool. Anz. 4: 285.
1881 ? *Dendrobaena puter* (part.)–Örley, Math. Term. Közlem. Magyar Akad. 16: 586.
1884 *Allolobophora constricta* Rosa, Lumbric. Piemonte, p. 38.

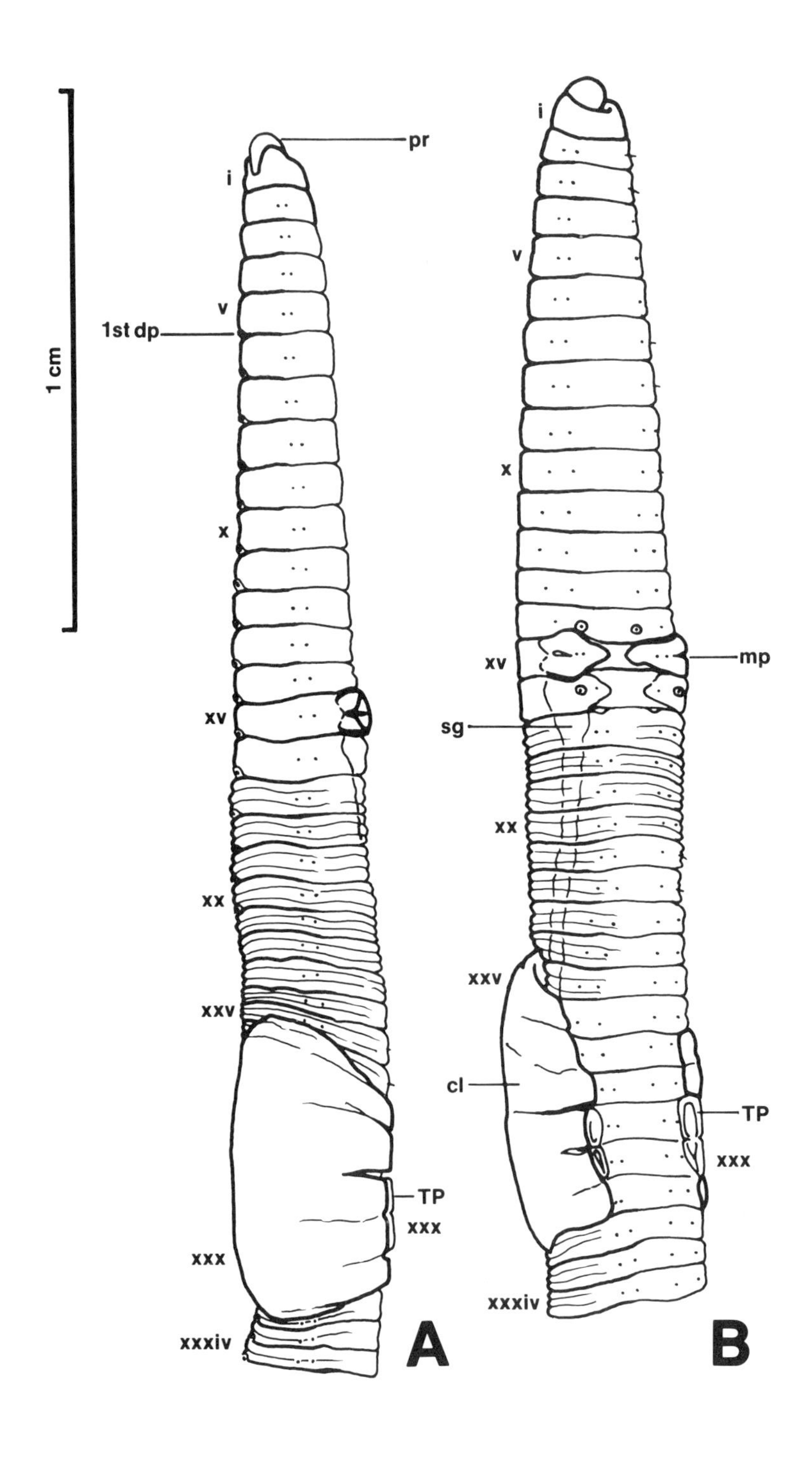

Fig. 20 External longitudinal views of *Dendrodrilus rubidus* showing taxonomic characters. A. Dorsolateral view. B. Ventrolateral view. (ONT: Manitoulin Dist., cat. no. 7502)

1884 *Lumbricus subrubicunda* (part.)–Levinsen, Vidensk. Meddel. Naturhist. Forh. Copenhagen, ser. 4, 5: 242.
1885 *Octolasion constrictum* + *O. subrubicundum*–Örley, Ertek. Term. Magyar Akad. 15(18): 20, 21.
1888 *Allolobophora putra* (part.)–Vejdovský, Entwickgesch. Unters., p. 41.
1889 *Lumbricus (Allolobophora) constrictus*–L. Vaillant, Hist. Nat. Annel. 3(1): 113.
1891 *Allolobophora nordenskioldii* (laps.) Michaelsen, Abh. Ver. Hamburg 11(2): 3.
1891 *Allolobophora rubicunda* (laps.)-Beddard, Proc. R. Phy. Soc. Edinburgh 10: 273.
1893 *Allolobophora putris arborea* (part. ?)–Rosa, Mem. Acc. Torino, ser. 2, 43: 433.
1893 *Dendrobaena constricta* (part.)–Friend, Naturalist, p. 19.
1896 *Allolobophora helvetica* + *A. darwini* Ribaucourt, Rev. Suisse Zool. 4: 18, 82.
1900 *Allolobophora (Bimastus) constricta*–Michaelsen, Abh. Nat. Ver. Hamburg 16(1): 10.
1900 *Helodrilus (Dendrobaena) rubidus*–Michaelsen, Das Tierreich, Oligochaeta 10: 490.
1908 *Helodrilus (Bimastus) constrictus*–Michaelsen, Denskschr. Med.-Naturew. Ges. Jena 13: 41.
1917 *Helodrilus (Bimastus) tenuis*–Smith, Proc. U.S. Natn. Mus. 52: 157, 182.
1958 *Dendrobaena rivulicola* Chandebois, Bull. Soc. Zool. France 83: 159.
1969 *Dendrobaena rubida* + *D. subrubicunda* + *D. tenuis*–Støp-Bowitz, Nytt. Mag. Zool. 17(2): 220, 224, 227.
1970 *Dendrobaena rubida* var. *typica*–Zajonc, Biol. Práce 16(8): 22.
1972 *Dendrobaena (Dendrodrilus) rubida rubida* + *D. (D.) rubida tenuis* + *D. (D.) subrubicunda*–Bouché, Inst. Natn. Rech. Agron., p. 410, 411, 414.
1972 *Dendrobaena subrubicunda* + *Bimastos tenuis* + *Dendrobaena rubida*–Edwards and Lofty, Biol. earthworms, p. 215, 216.
1973 *Dendrobaena (Dendrodrilus) rubida* + *D. (D.) r.* f. *typica* + *D. (D.) r.* f. *subrubicunda* + *D. (D.) r.* f. *tenuis*–Plisko, Fauna Polski, no. 1, p. 79, 84, 85, 87.
1975 *Dendrodrilus rubidus*–Reynolds, Megadrilogica 2(3): 3.

Diagnosis

Length 20–90 mm, diameter 2–5 mm, segment number 50–120, prostomium epilobic, first dorsal pore 5/6. Clitellum xxvi, xxvii–xxxi, xxxii. Tubercula pubertatis, if present, xxviii, xxix–xxx. Setae widely paired, $AB < CD$, and $BC = 2\ CD$. Male pores on xv between *b* and *c*. Seminal vesicles in 9, 11, and 12. Spermathecae, two pairs with short ducts on level with setae *c* opening in 9/10 and 10/11. Body cylindrical. Colour, red and darker dorsally.

Biology

Dendrodrilus rubidus has been found in a wide range of habitats including gardens, cultivated fields, stream banks, in moss in running water and wells and springs, peat, compost, and sometimes in manure. It seems acid tolerant. The species is known from caves in Europe and North America, and in greenhouses, botanical gardens, and the culture beds of earthworm farms (Gates, 1972c). *Dd. rubidus* lives in the upper soil layers though on wet nights the worms have been seen wandering on the soil surface or climbing trees. Under experimental conditions it can withstand prolonged immersion in water (Roots, 1956). Černosvitov and Evans (1947) and Gerard (1964) reported it from under the bark of old trees, and under moss, leaf mould, or rotten wood in moist areas. In Ontario *Dd. rubidus* was most frequently found under logs.

Although activity can be year round it is probable that in much of North America, including Ontario, a winter rest is imposed by the climate. Copulation has not been properly studied but one published observation records an unusual position of ventral apposition, head to head and tail to tail (see Gates, 1972c). *Dd. rubidus* is facultatively parthenogenetic with male sterility and absence of spermathecae common (Gates, 1972c; Reynolds, 1974c).

Range

A native of Palaearctis, *Dd. rubidus* is now know from Europe, North America, South America, Asia, Africa, and Australasia (Gates, 1972c). Also known from Iceland (Backlund, 1949).

North American Distribution

Alberta (Gates, 1972c), British Columbia (Gates, 1972c), Manitoba (Gates, 1972c), New Brunswick (Reynolds, 1976d), Newfoundland (Eaton, 1942), Nova Scotia (Reynolds, 1975a, 1976a), Ontario (Eisen, 1874), Prince Edward Island (Reynolds, 1975c), Québec (Reynolds, 1975d, 1976c), Saskatchewan (Gates, 1972c), Alabama (Smith, 1917), Alaska (Michaelsen, 1903a), Arizona (Reynolds et al., 1974), Arkansas (Causey, 1953), California (Smith, 1917), Connecticut (Reynolds, 1973c), Delaware (Reynolds, 1973a), Hawaii (Ude, 1905), Idaho (Gates, 1967), Illinois (Smith 1917), Indiana (Heimburger, 1915), Kentucky (Giavannoli, 1933), Louisiana (Harman, 1952), Maine (Gates, 1949), Maryland (Reynolds, 1974b), Massachusetts (Reynolds, 1977), Michigan (Smith, 1917), Missouri (Olson, 1936), Montana (Reynolds, 1972c), Nevada (Gates, 1967), New Hampshire (Southern, 1910), New Jersey (Eaton, 1942), New York (Olson, 1940), Ohio (Olson, 1928), Oregon (Gates, 1972c), Pennsylvania (Gates, 1959), Rhode Island (Reynolds, 1973a), Tennessee (Reynolds, 1974a), Utah (Gates, 1967), Vermont (Reynolds, 1976c), Virginia (Reynolds, 1974b), Washington (Smith, 1917), West Virginia (Gates, 1972c), Wyoming (Gates, 1967), Greenland (Omodeo, 1955a). New records: Florida, Georgia, Iowa, Minnesota, New Mexico, North Carolina.

Ontario Distribution (Fig. 21)

Dendrodrilus rubidus was the third species reported for Ontario by Eisen (1874). Later reports of this species in Ontario were made by Stafford (1902), Gates (1943), Judd (1964), and Reynolds (1972a).

ALGOMA DIST. *Hwy 17, 7.58 km e of Spanish, under logs, 13 May 72, JWR & JEM, 0-0-1. Alona Bay Mine, 4 Jun 71, GM, 0-1-0. Wawa Mine, Wawa, 5 Jun 71, GM, 0-1-0. BRANT CO. *Hwy 54, .64 km w of Onondaga, under logs in wet area, 3 May 72, JWR, 0-1-3. BRUCE CO. *Hwy 4, 6.77 km s of Teeswater, under logs, 5 May 72, JWR, 0-0-1. Hwy 21, 2.57 km n of Tiverton, under lumber, 5 May 72, JWR, 0-1-1. *Hwy 21, 1.61 km e of Elsinore, under logs, 5 May 72, JWR, 0-1-1. COCHRANE DIST. Smoky Falls, under lumber, 16 Apr 38, RVW, 0-0-2, ROM-126. DUFFERIN CO. *Hwy 10-24, 8.06 km n of Orangeville, under logs, 2 May 72, JWR, 0-1-6. DURHAM CO. *Hwy 401, .32 km e of Mill St, Newcastle, under logs, 15 May 72, JWR, 0-0-1. ESSEX CO. *Hwy 3, Ruthven, n.e., wet ditch next to railroad tracks, 4 May 72, JWR, 0-1-0. *Hwy 2, 8.55 km e of St. Joachim, ditch, 4 May 72, JWR, 0-1-0. FRONTENAC CO. *Hwy 2, .97 km w of Westbrook, under logs, 16 May 72, JWR, 0-0-1. *Hwy 7, 15.81 km e of Kaladar, under logs, 16 May 72, JWR, 0-0-1. GRENVILLE CO. Hwy 43, Merrickville, in grass clippings, 11 May 72, JWR, 0-0-2. GREY CO.

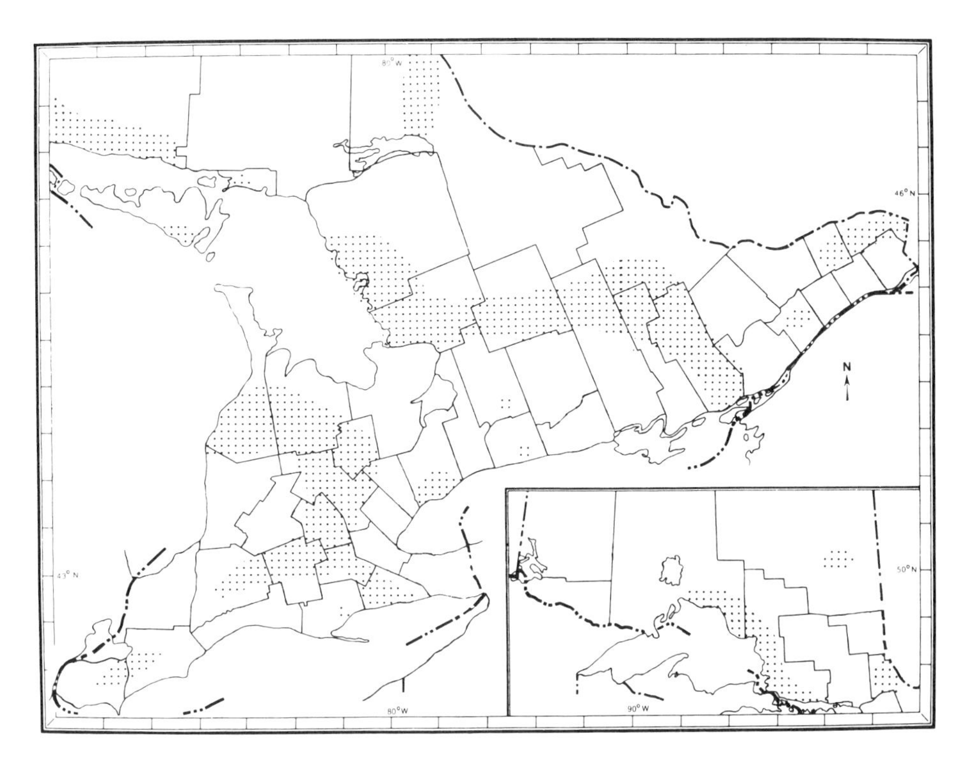

Fig. 21 The known Ontario distribution of *Dendrodrilus rubidus.*

*Hwy 6BP, Rockford, w.e., under logs, 5 May 72, JWR 1-0-1. *Hwy 4, 8.22 km e of Durham, under logs, 5 May 72, JWR, 0-0-2. HALDIMAND CO. *Hwy 3, 6.29 km w of Dunnville, under logs, 1 May 72, JWR, 0-0-1. HALIBURTON CO. Reynolds (1972a). HASTINGS CO. *Hwy 62, .64 km w of Maple Leaf, under log, 16 May 72, JWR, 1-1-3. KENT CO. *River Rd, 6.45 km w of Chatham, under log, 23 Apr 72, JWR & TW, 0-0-1. *Hwy 3, 9.03 km e of Port Alma, under log, 4 May 72, JWR, 0-1-0. MANITOULIN DIST. Gates (1943). *Hwy 68, 3.22 km n of little Current, under logs and lumber, 13 May 72, JWR & JEM, 0-0-2. MIDDLESEX CO. Judd (1964). MUSKOKA DIST. *Hwy 11, 1.13 km n of Severn Bridge, under log, 9 May 72, JWR, 0-0-1. *Hwy 103, 2.26 km n of Hwy 660, under logs in ditch, 9 May 72, JWR, 0-0-3. NIAGARA CO. Niagara, 1873, GE, 0-0-1, USNM-18937. NIPISSING DIST. Hwy 17, 15.81 km e of Sturgeon Falls, dump, 13 May 72, JWR & JEM, 0-0-2. Lake Temagami, under rotten logs, 2 Jul 37, GMN, 1-0-2, ROM-I24. NORFOLK CO. St. Williams Forestry Experimental Station, 5 Oct 75, DPS, 0-0-1, UW-0001. ONTARIO CO. Dufferin Creek area, day after use of lampricide, 11 May 71, TY, 0-0-1, ROM-I39. OXFORD CO. *Hwy 59, 1.94 km s of Norwich, under log, 1 May 72, JWR, 0-0-1. PARRY SOUND DIST. *Hwy 124, 3.06 km e of McKellar, under logs, 9 May 72, JWR, 1-1-1. PRESCOTT CO. *Hwy 34, 5.81 km n of Vankleek Hill, under logs, 11 May 72, JWR, 0-2-6. Hwy 17, .64 km e of Plantagenet, under log, 11 May 72, JWR, 0-0-1. RENFREW CO. Reynolds (1972a). *2.26 km n of Schutt, ditch, 16 May 72, JWR, 0-0-1. *4.19 km s of Palmer Rapids, under paper in wet ditch, 16 May 72, JWR, 0-0-3. *2.42 km s of Palmer Rapids, under paper in ditch, 16 May 72, JWR, 0-0-3. RUSSELL CO. *2.96 km e of Embrun, under logs, 11 May 72, JWR, 0-0-1. SUDBURY DIST. Webbwood Mine, Webbwood, 1 Jun 71, GM, 1-1-0. THUNDER BAY DIST. Jackfish Mine, Terrace Bay, 6 Jun 71, GM, 0-1-1. Schrieber Mine, 6 June 71, GM, 0-1-2. VICTORIA CO. *Hwy 7, 7.26 km e of Hwy 35, under log, 26 Apr 72, JWR, 0-0-1. WATERLOO CO. *Hwy 97, Galt, .81 km w of Hwy 24A, under grass clippings, 3 May 72, JWR, 9-0-9. Waterloo, Laurel Creek Conservation Area, 3 Aug 74, DPS, 0-3-5, UW-0007. WELLINGTON CO. *Hwy 6, 1.4 km n of Aberfoyle, under logs, 29 Apr 71, JWR, 0-0-3. *Hwy 6, 10.64 km s of Arthur, under logs, 2 May 72, JWR, 0-0-2. YORK CO. *Hwy 48, .32 km n of Steeles Avenue, under logs, 26 Apr 72, JWR, 1-0-8. Hwy 27, 8.55 km s of Hwy 9, under log, 29 Apr 72, JWR, 0-0-1. Edenbrook Park, Islington, quantitative study 2, formalin, 18 May 72, JWR, 0-1-3.

Genus *Eisenia* Malm, 1877

1877 *Eisenia* Malm, Öfv. Salsk. Hortik. Förh. Göteborg 1: 45.
1900 *Eisenia* (part.)–Michaelsen, Das Tierreich, Oligochaeta 10: 474.
1969 *Eisenia*–Gates, J. Nat. Hist. Lond. 9: 305.

Type Species
Enterion foetidum Savigny, 1826 by Gates (1969).

Diagnosis
Calciferous gland, without sacs, opening into gut behind insertion of 10/11 through a circumferential circle of small pores. Calciferous sacs, lacking. Gizzard, mostly in xvii. Hearts, in vii–xi. Nephridial bladders, sausage-shaped or digitiform, transversely placed. Nephropores, inconspicuous, in two ranks on each side, alternating irregularly and with asymmetry between a level just above *B* and one above *D*. Setae, closely paired. Prostomium epolobic. Longitudinal musculature, pinnate. Pigment, red. (after Gates, 1972c: 96; 1975a: 3).

Discussion
Eisenia was erected for three species, *Enterion foetidum* Savigny, 1826, and *Allolobophora norvegica* and *A. subrubicunda* Eisen, 1873 by Malm (1877) without the designation of a type species. The two *Allolobophora* species are now synonyms of *Dendrodrilus rubidus*. Gates (1969) redefined *Eisenia* with *Eisenia foetida* as type species, but another of Savigny's species, *Enterion roseum*, has been congeneric with *Eisenia foetida* since Michaelsen (1900b) solely because of spermathecal pore location. If future revisions are based on the more conservative somatic anatomy these two species will not remain congeneric. Most European workers have followed Pop (1941) and Omodeo (1956) and have transferred *E. rosea* to *Allolobophora*. However, on the basis of somatic anatomy it is not reasonable to place *E. rosea* in any genus of which *Enterion chloroticum* Savigny is the type species. Until this problem is solved I will continue to use *Eisenia rosea* to reduce confusion. Recently Gates (1976) transferred *E. rosea* to the genus *Aporrectodea*.

Eisenia foetida (Savigny, 1826)
Manure worm Ver du fumier
(Fig. 22)

1826 *Enterion fetidum* (corr. *foetidum*) Savigny, Mém. Acad. Sci. Inst. Fr. 5: 182.
1835 *Lumbricus semifasciatus* Burmeister, Zool. Hand. Atl. 33: 3.
1836 *Lumbricus annularis* R. Templeton, Ann. Mag. Nat. Hist. 9: 234.
1837 *Lumbricus foetidus*–Dugès, Ann. Sci. Nat., ser. 2, 8: 17, 21.
1842 *Lumbricus olidus* Hoffmeister, Verm. Lumbric., p. 25.
1849 ? *Lumbricus luteus* Blanchard, Hist. Chile 3: 42.

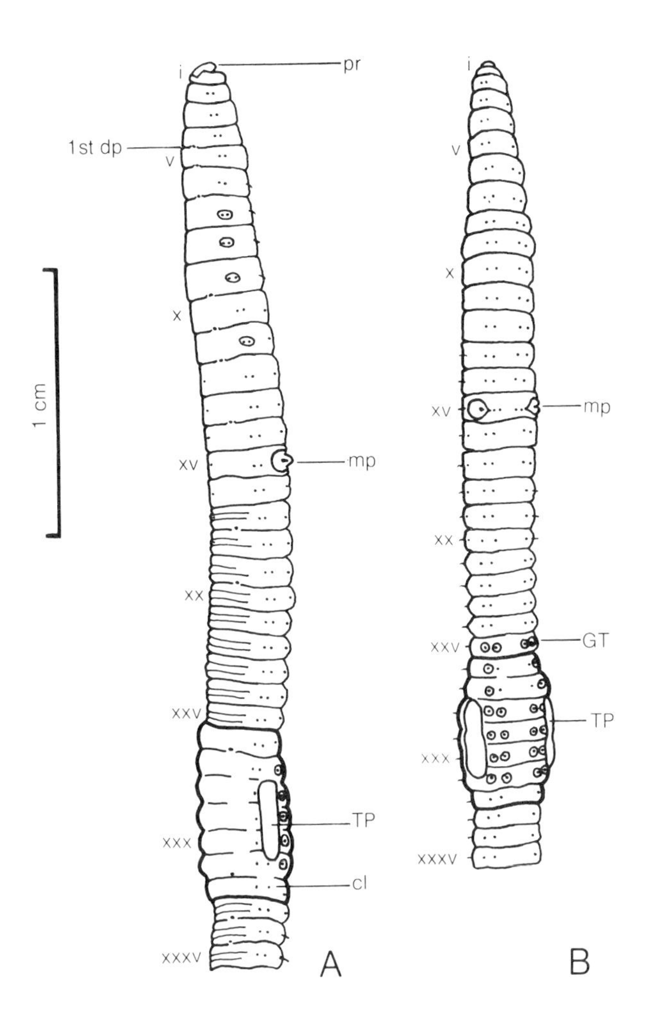

Fig. 22 External longitudinal views of *Eisenia foetida* showing taxonomic characters. A. Lateral view. B. Ventral view. (ONT: Nipissing Dist., cat. no. 7604)

1873 ? *Lumbricus rubro-fasciatus* Baird, Proc. Linn. Soc. Lond. 11: 96.
1873 *Allolobophora foetida*–Eisen, Öfv. Vet.-Akad. Förh. Stockholm 30(8): 50.
1877 *Lumbricus annulatus* Hutton, Trans. N.Z. Inst. 9: 352.
1877 *Eisenia foetida*–Malm, Öfv. Salsk. Hortik. Förh. Göteborg 1: 45.
1887 *Endrilus annulatus* W.W. Smith, Trans. N.Z. Inst. 19: 136.
1889 *Lumbricus (Allobophora) annulatus* + *L. (A.) foetidus*–L. Vaillant, Hist. Nat. Annel. 3(1): 147, 149.
1913 *Helodrilus (Eisenia) foetidus*–Michaelsen, Zool. Jb. Syst. 34: 551.
1963 *Eisenia foetida* var. *unicolor* André, Bull. Biol. Fr. Belg. 81: 1.
1972 *Eisenia fetida fetida* + *E. f. andrei* Bouché, Inst. Natn. Rech. Agron., p. 380, 381.

Diagnosis
Length 35–130 mm (generally < 70 mm), diameter 3–5 mm, segment number 80–110, prostomium epilobic, first dorsal pore 4/5 (sometimes 5/6). Clitellum xxiv, xxv, xxvi–xxxii. Tubercula pubertatis on xxviii–xxx. Setae closely paired, $AB = CD$, $BC < AA$, anteriorly DD = ½C but posteriorly DD<½C. Genital tumescences may be present around any of the setae on ix–xii, usually around setae *a* and *b* of xxiv–xxxii. Male pores with large glandular papillae on segment xv. Seminal vesicles, four pairs in 9–12. Spermathecae, two pairs with ducts opening near mD line in 9/10 and 10/11. Colour variable, purple, red, dark red, brownish red, sometimes alternating bands of red-brown on dorsum with pigmentless yellow intersegmental areas. Body cylindrical.

Biology
Olson (1928) found this species in manure and decaying vegetation where moisture concentrations were high. Černosvitov and Evans (1947) and Gerard (1964) recorded its habitats as manure, compost heaps, and soil high in organic matter, as well as forests, gardens, and under stones and leaves. Murchie (1956) reported *E. foetida* from manure and bait castaways but never from what he considered "natural" habitats. In Tennessee Reynolds et al. (1974) recorded scattered distribution of this species with it occurring most commonly under logs and debris and at roadside dumps. According to Gates (1972c) the available records give a pH range of 6.8–7.6, and while in Scandinavia it has been considered a species dependent on human culture, *E. foetida* is known from caves in Europe and North America, and Russian records report it from taiga, forests, and steppes. In Ontario *E. foetida* was found most frequently under logs and usually not far from human habitation. There are few data available concerning its natural habitat in North America.

There is little information on rest periods in the life-cycle (Gates, 1972c). One assumes that under favourable conditions activity can occur throughout the year. Feeding is selective in that there is minimal ingestion of earth. Copulation is subterranean and although the species has been thought to be obligatorily amphimictic, uniparental reproduction is possible, though very rare (Gates, 1972c). Experimental self-fertilizing was demonstrated by André (1963). *E. foetida* has a maximum life expectancy of 4–5 years, although between 1 and 2 years is more usual.

Eisenia foetida has been reared on earthworm farms and sold in every Canadian province and American state for fish bait. Harman (1955) reported on the commercial aspects of this species.

Range
A native of Palaearctis, *E. foetida* is now known from Europe, North America, South America, Asia, Africa, and Australasia (Gates, 1972c). Also known from Iceland (Backlund, 1949).

North American Distribution

British Columbia (Gates, 1942), Nova Scotia (Reynolds, 1975a, 1976a), Ontario (Reynolds, 1972a), Québec (Reynolds, 1975e, 1976c), Alabama (Gates, 1972c), Arizona (Gates, 1967), Arkansas (Causey, 1952), California (Gates, 1943), Connecticut (Reynolds, 1973c), District of Columbia (Gates,

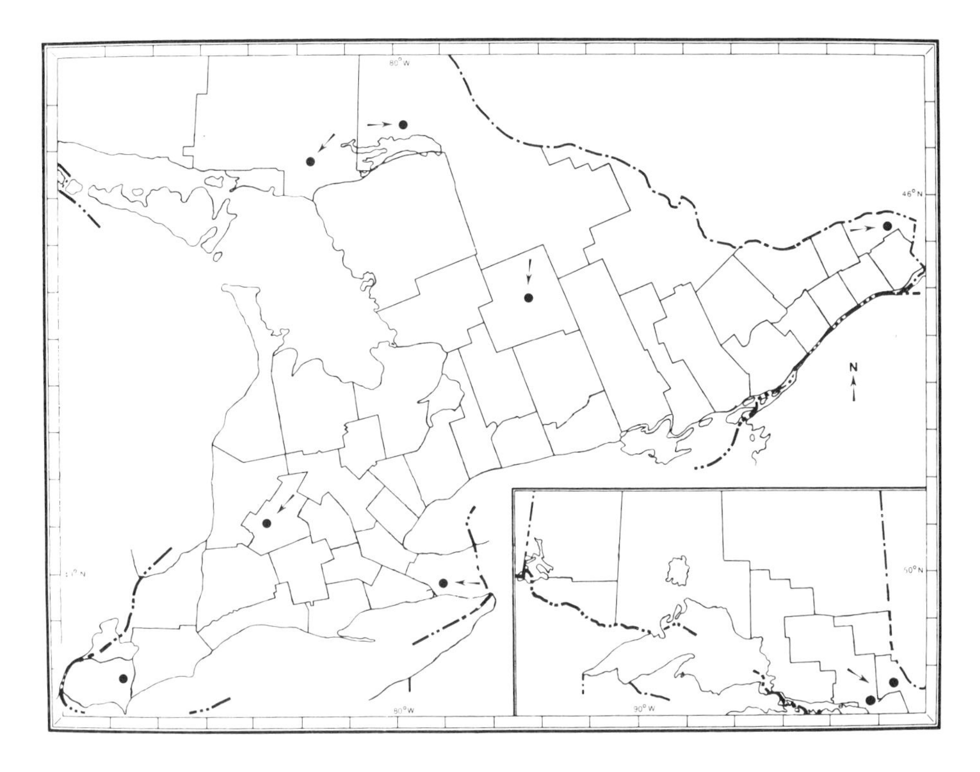

Fig. 23 The known Ontario distribution of *Eisenia foetida*.

1972c), Florida (Harman, 1955), Georgia (Harman, 1955), Hawaii (Michaelsen, 1900b), Illinois (Smith, 1928), Indiana (Heimburger, 1915), Iowa (Harman, 1955), Kentucky (Giavannoli, 1933), Louisiana (Harman, 1952), Maine (Gates, 1966), Maryland (Reynolds, 1974b), Massachusetts (Reynolds, 1977), Michigan (Smith and Green, 1916), Minnesota (Mickel, 1925), Mississippi (Harman, 1955), Missouri (Olson, 1936), Nebraska (Swartz, 1929), New Hampshire (Gates, 1972c), New Jersey (Eaton, 1942), New York (Olson, 1940), North Dakota (Harman, 1955), Ohio (Olson, 1928), Oklahoma (Harman, 1954), Oregon (MacNab and McKey-Fender, 1947), Pennsylvania (Andrews, 1895), Rhode Island (Reynolds, 1973b), South Carolina (Harman, 1955), South Dakota (Gates, 1967), Tennessee (Reynolds, 1974a), Texas (Harman, 1955), Utah (Gates, 1967), Vermont (Reynolds, 1976c), Virginia (Gates, 1949), Washington (Altman, 1936), West Virginia (Gates, 1967), Greenland (Michaelsen, 1892). New record: New Mexico.

Ontario Distribution (Fig. 22)

First reported by Stafford (1902), and suspected from Haliburton County (Reynolds, 1972a), *Eisenia foetida* is not widespread in Ontario.

ESSEX CO. *Hwy 3, 1.45 km e of Cottam, under logs, 4 May 72, JWR, 0-0-2. HALIBURTON CO. *Reynolds (1972a). KENT CO. *River Rd. 6.45 km w of Chatham, compost pile, 23 Apr 72, JWR & TW, 2-1-2. NIAGARA CO. *Hwy 20, 1.61 km w of Smithville, old house site, 1 May 72, JWR, 0-0-1. NIPISSING DIST. *Hwy 17, Sturgeon Falls, w.e., under logs, 13 May 72, JWR & JEM, 0-1-5. PERTH CO. *Mitchell, next to Collegiate, under log in farmyard, 4 May 72, JWR & DWR, 0-0-1. PRESCOTT CO. *Hwy 34, 2.58 km n of Vankleek Hill, under logs and rocks, 11 May 72, JWR, 4-8-1. SUDBURY DIST. *Hwy 69, 6.61 km s of Hwy 637, under grass clippings, 13 May 72, JWR & JEM, 0-2-0.

***Eisenia rosea* (Savigny, 1826)**

Pink soil worm Ver rose du sol

(Fig. 24)

1826 *Enterion roseum* Savigny, Mém. Acad. Sci. Inst. Fr. 5: 182.
1837 *Lumbricus roseus*–Dugès, Ann. Sci. Nat., ser. 2, 8: 17, 20.
1845 *Lumbricus communis anatomicus* (part.) Hoffmeister, Regenwürmer, p. 28.
1873 *Allolobophora mucosa* (part.) Eisen, Öfv. Vet.-Akad. Förh. Stockholm 30(8): 47.
1875 *Lumbricus aquatilis* Vejdovský, SB Bohm. Ges., p. 199.
1879 *Lumbricus muscosus*–Tauber, Annul. Danmark, p. 68.
1882 *Lumbricus carneus* (err. *non Enterion carneum* Savigny, 1826)–Vejdovský, Brunnenw. Prague, p. 51.
1884 *Allolobophora carnea*–Vejdovský, Syst. Morph. Oligochäten, p. 61.
1885 *Allolobophora aquatilis* + *A. aguatilis*–Örley, Ertek. Term. Magyar Akad. 15(18): 24, 28.
1893 *Allolobophora rosea*–Rosa, Mem. Acc. Torino, ser. 2, 43: 424, 427.
1896 *Allolobophora danieli rosai*–Ribaucourt, Rev. Suisse Zool. 4: 39.
1900 *Eisenia rosea*–Michaelsen, Das Tierreich, Oligochaeta 10: 478.
1903 *Helodrilus (Bimastus) bimastoides* Michaelsen, Mitt. Naturh. Mus. Hamburg 19: 13.
1907 *Helodrilus (Bimastus) indicus* Michaelsen, Mitt. Naturh. Mus. Hamburg 24: 188.
1917 *Helodrilus (Eisenia) roseus*–Smith, Proc. U.S. Natn. Mus. 52(2174): 165, 166.
1922 *Helodrilus (Allolobophora) prashadi* Stephenson + *A. (Bimastus) indica*–Stephenson, Rec. Indian Mus. 22: 440, 441.
1940 *Eisenia rosea* f. *typica* + *E. r.* f. *macedonica* + *Allolobophora hataii* + *A. harbinensis* + *A. dairensis* + *A. jeholensis* Kobayashi, Sci. Rep. Tôhoku Imp. Univ., ser. 4, 15: 285-287, 288-289, 290-291, 291-293, 293-295.
1949 *Eophila kulagini* Malevič, Dokl. Akad. Nauk SSSR (Biol.) 47: 400.
1967 *Allolobophora rosea* var. *alpina* Vedovini, Bull. Soc. Zool. Fr. 92: 793.
1970 *Allolobophora rosea*–Zajonc, Biol. Práce 16(8): 23.
1972 *Allolobophora rosea*–Edwards and Lofty, Biol. earthworms, p. 217.
1972 *Allolobophora rosea rosea*–Bouché + *A. r. vedovinii* Bouché, Inst. Natn. Rech. Agron., p. 418, 423.
1976 *Aporrectodea rosea*–Gates, Megadrilogica 2(12): 4.

Diagnosis

Length 25–85 mm, diameter 3–5 mm, segment number 120–150, prostomium epilobic, first dorsal pore 4/5. Clitellum, somewhat flared ventrally xxv, xxvi–xxxii. Tubercula pubertatis usually xxix–xxxi. Setae closely paired, $AA>BC<DD$, $AB>CD$, anteriorly DD = ½C, posteriorly DD = ⅓C. Male pores with elevated glandular papillae in xv with male tumescences extending over xiv and xvi. Seminal vesicles, four pairs in 9–12. Spermathecae, two pairs with short ducts opening near mD line or halfway between mL and *d* in 9/10 and 10/11. Body cylindrical, except in clitellar region. Unpigmented, but colour

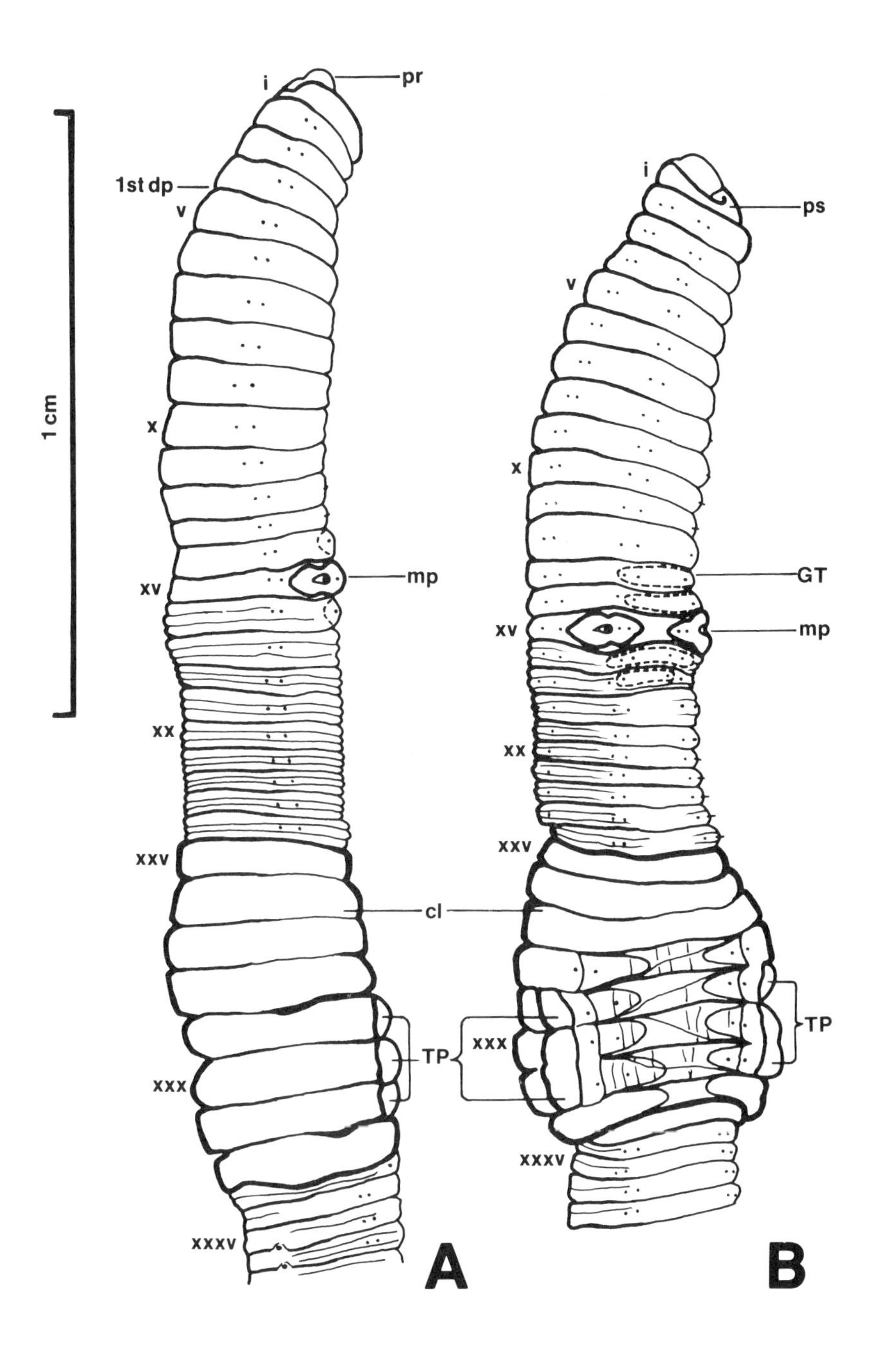

Fig. 24 External longitudinal views of *Eisenia rosea* showing taxonomic characters. A. Dorsolateral view. (ONT: Perth Co., cat. no. 7664) B. Ventrolateral view. (ONT: Nipissing Dist., cat. no. 7535)

appears rosy or greyish when alive, and white when preserved.

Biology

Černosvitov and Evans (1947) and Gerard (1964) recorded *E. rosea* in soil, fields, gardens, pastures, and forests, under leaves and stones, and frequently on river and lake banks. Gates (1972c) mentioned soils of pH 4.9–8.0 and recorded that it is found often enough in conditions that were thought to justify characterization as "amphibious". Murchie (1956) states, "*Eisenia rosea*, although showing considerable adaptability in habitat requirements, is to be regarded primarily as a true soil species". The results of this survey would tend to confirm Murchie's statement. In soil under logs was the most common habitat of this species in Ontario. *E. rosea* is one of the cosmopolitan species that have been introduced by Europeans into all parts of the world. It is known also from caves in Europe, Asia, and North America, as well as from botanical gardens and greenhouses. It is the only species widely distributed in the virgin steppes of Russia and the mountains of the Caucasus (Gates, 1972c).

In suitable conditions activity, including breeding, is possible the year round. But in northern parts of the range there is a resting stage during winter cold and summer drought in which both hibernation and aestivation are spent tightly coiled in a small pink ball (Gates, 1972c). According to Thomson and Davies (1974), *E. rosea* produces surface casts, despite some contrary statements in the literature. The species is parthenogenetic and biparental reproduction of anthropochorous morphs is unknown (Gates, 1974c, Reynolds, 1974c). Černosvitov (1930) reported degeneration, phagocytosis, and reabsorption of sperm and Tuzet (1946) recorded atypical spermatogenesis. Evans and Guild (1948) reared isolated individuals to sexual maturity which then produced fertile cocoons.

Eisenia rosea is the primary host of the cluster fly, *Pollenia rudis* (Fabr.) (Yahnke and George, 1972; Thomson and Davies, 1973a).

Range

A native of Palaearctis, *E. rosea* is now known from Europe, North America, South America, Africa, Asia, and Australasia. It also occurs in Iceland (Backlund, 1949). Generally, therefore, it is a cosmopolite although apparently absent from tropical lowlands (Gates, 1972c).

North American Distribution

Alberta (Gates, 1972c), British Columbia (Gates, 1974c), Labrador (Gates, 1974c), New Brunswick (Reynolds, 1976d), Nova Scotia (Reynolds, 1975a, 1976a), Ontario (Reynolds, 1972a), Prince Edward Island (Reynolds, 1975c), Québec (Reynolds, 1975d, e, 1976c), Arizona (Smith, 1917), Arkansas (Causey, 1952), California (Smith, 1917), Colorado (Gates, 1967), Connecticut (Reynolds, 1973c), Delaware (Reynolds, 1973a), Florida (Gates, 1974c), Georgia (Smith, 1917), Hawaii (Gates, 1972c), Idaho (Gates, 1967), Illinois (Smith, 1917), Indiana (Heimburger, 1915), Iowa (Gates, 1967), Kansas (Gates, 1974c), Kentucky (Gates, 1959), Louisiana (Smith, 1917), Maine (Smith, 1917), Maryland (Reynolds, 1974b), Massachusetts (Reynolds, 1977), Michigan (Murchie, 1956), Minnesota (Gates, 1974c), Missouri (Olson, 1936), Montana (Reynolds, 1972c), Nevada (Gates, 1967), New Hampshire (Gates, 1972c), New Jersey (Davies, 1954), New Mexico (Gates, 1967), New York (Smith, 1917), North Carolina (Pearse, 1946), Ohio (Olson, 1928), Oklahoma (Harman, 1954), Oregon (MacNab and McKey-Fender, 1947), Pennsylvania (Bhatti, 1965), Rhode Island (Reynolds, 1973b), Tennessee (Reynolds, 1974a), Texas (Gates, 1972c), Utah (Gates, 1972c), Vermont (Gates, 1949), Virginia (Gates, 1959), Washington (Gates, 1972c), West Virginia (Reynolds, 1974b), Wyoming (Gates, 1967).

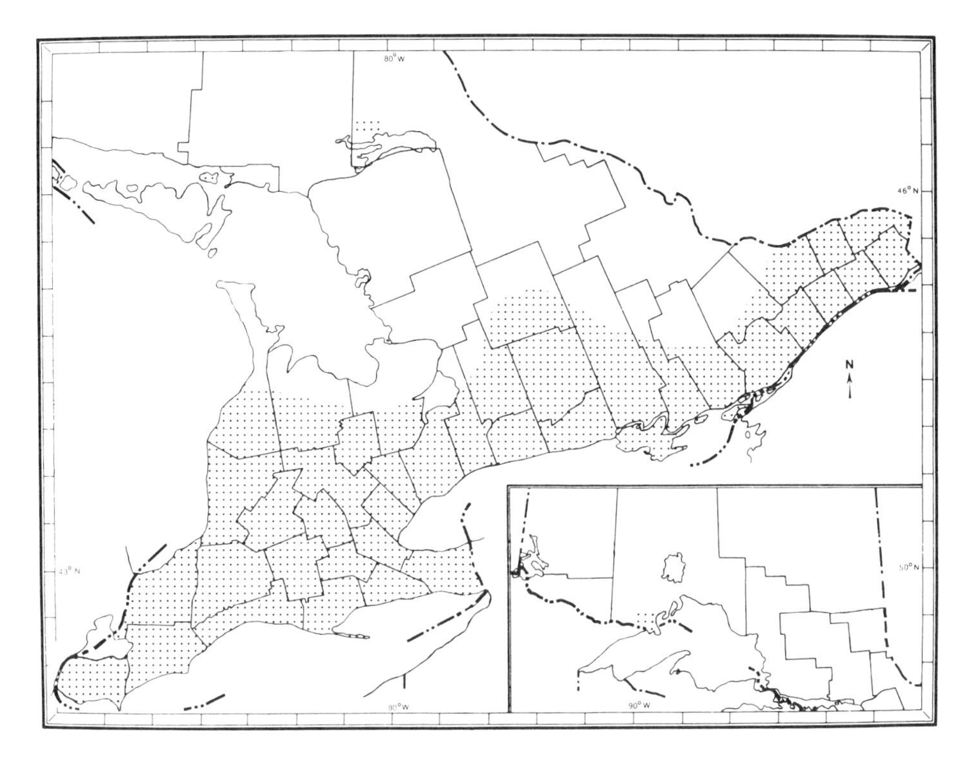

Fig. 25 The known Ontario distribution of *Eisenia rosea.*

Ontario Distribution (Fig. 25)

This species was first recorded from Ontario by Stafford (1902) and then by Judd (1964) and Reynolds (1972a). In the present survey it was found in all but four counties in southern Ontario, but additional collecting under favourable conditions should produce specimens from these counties.

BRANT CO. *Hwy 53, 5.48 km w of Cathcart, under railroad ties, 1 May 72, JWR, 1-1-3. *Hwy 54, .64 km w of Onondaga, under log in wet area, 3 May 72, JWR, 0-0-1. BRUCE CO. *Hwy 4, 6.77 km s of Teeswater, under logs, 5 May 72, JWR, 4-1-1. Hwy 21, 2.57 km n of Tiverton, under lumber, 5 May 72, JWR, 0-6-2. *Hwy 21, 1.61 km e of Elsinore, under logs, 5 May 72, JWR, 3-0-0-1. CARLETON CO. Reynolds (1972a). *Ottawa-Carleton Rd 35, 2.42 km s of Leonard, under logs, 11 May 72, JWR, 3-1-3. DUFFERIN CO. Hwy 9, 1.61 km e of Hwy 10, under logs, 29 Apr 72, JWR, 0-0-6. *Hwy 9, 2.1 km w of Hwy 104, digging in road bank, 2 May 72, JWR, 0-0-1. *Hwy 9, 3.71 km w of Orangeville, under logs in ditch, 2 May 72, JWR, 0-2-1. *Hwy 10-24, 8.06 km n of Orangeville, under logs, 2 May 72, JWR, 3-0-3. DUNDAS CO. *Hwy 2, 5 km w of Iroquois, under log, 11 May 72, JWR, 5-3-5. *Hwy 2, 5.48 km e of Iroquois, under logs, 11 May 72, JWR, 0-0-15. *Hwy 31, 3.55 km n of Morrisburg, under lumber, 11 May 72, JWR, 0-0-2. DURHAM CO. Hwy 7A, 2.1 km e of Nestletown Station, under logs, 26 Apr 72, JWR, 0-2-1. *Hwy 401, .61 km w of Liberty Rd, Bowmanville, digging under log next to railroad tracks, 15 May 72, JWR, 2-0-3. *Hwy 402, .32 km e of Mill St, Newcastle, under logs, 15 May 72, JWR, 0-0-2. *Hwy 402, 1.61 km e of Newtonville, under log, 15 May 72, JWR, 0-0-1. ELGIN CO. *Hwy 73, 3.06 km s of Harrietsville, under rotten log, 4 May 72, JWR, 3-1-6. *Hwy 73, 6.77 km n of Aylmer, under logs, 4 May 72, JWR, 0-0-6. Hwy 3, 9.19 km s of St. Thomas, under logs, 4 May 72, JWR, 0-2-8. Hwy 3, 5.16 km w of Talbotville, under log, 4 May 72, JWR, 5-0-6. *Hwy 6.77 km e of Wallacetown, under pine logs, 4 May 72, JWR, 2-2-0. Hwy 3, 7.9 km w of Frome, under logs, 4 May 72, JWR, 1-2-3. ESSEX CO. *Hwy 3, 1.45 km e of Cottam, under logs, 4 May 72, JWR, 0-0-4. *Belle River, s.e., under logs in dump, 4 May 72, JWR, 1-0-1. FRONTENAC CO. *Hwy 15, 4.68 km s of Seeley's Bay, under logs, 16 May 72, JWR, 0-0-3. GLENGARRY CO. *Hwy 34, 11.13 km n of Lancaster, under log, 11 May 72, JWR, 2-1-4. *Hwy 34, 5.64 km n of Alexandria, under logs, 11 May 72, JWR, 3-0-13. GRENVILLE CO. *Hwy 2, Johnstown, w.e., under logs, 11 May 72, JWR, 5-0-6. Hwy 43, Merrickville, under log, 11 May 72, JWR, 0-0-1. GREY CO. *Hwy 6BP, Rockford, w.e., under log, 5 May 72, JWR, 0-0-1. *Hwy 6, 3.06 km s of Chatsworth, under logs, 5 May 72, JWR, 1-0-3. *Hwy 10, 8.06 km s of Flesherton, under logs, 5 May 72, JWR, 1-0-1. HALDIMAND CO. *Hwy 3, 6.13 km w of Cayuga, under log, 1 May 72, JWR, 0-0-1. *Hwy 54, 2.26 km w of Caledonia, digging, 3 May 72, JWR, 0-0-2. HALIBURTON CO. *Reynolds (1972a). HALTON CO. *Hwy 5, 9.19 km w of Hwy 10, under logs next to barn, 29 Apr 72, 1-1-0. *Halton Co. Rd 3, Esquesing Community, under rock, next to Town Hall, 4 May 73, JWR, 0-1-0. *Hwy 7, 4.84 km e of Georgetown, under burnt log in soil, 4 May 73, JWR, 0-0-1. HASTINGS CO. *Hwy 62, 1.61 km n of Hwy 7, under log in wet ditch, 27 Apr 72, JWR, 0-0-1. HURON CO. *Hwy 83, 6.77 km w of Russeldale, under logs, 5 May 72, JWR, 0-2-5-1. *Hwy 4, 3.71 km s of Brucefield, under logs and rocks, 5 May 73, JWR, 3-3-7. Hwy 4, 1.29 km n of Hensall, under corn (*Zea maize*) cobs in pasture, 5 May 72, JWR, 0-0-1. Hwy 4, 5.16 km s of Clinton, under logs, 5 May 72, JWR, 4-1-3. *Hwy 4, 2.74 km n of Clinton, under log, 5 May 72, JWR, 0-0-1. KENT CO. Hwy 3, 1.77 km w of Palmyra, under logs, 4 May 72, JWR, 1-1-2. *Hwy 3, 9.03 km e of Port Alma, under log, 4 May 72, JWR, 0-1-0. *Hwy 3, 3.55 km e of Wheatley, digging, 4 May 72, JWR, 4-1-1. LAMBTON CO. *Hwy 21, 6.45 km n of Dresden, under log, 4 May 72, JWR, 0-0-1. Hwy 21, 11.29 km n of Dresden, under dung in pasture, 4 May 72, JWR, 0-0-1. Hwy 21, .64 km s of Petrolia, ditch, 4 May 72, JWR, 3-0-1. *Hwy 21, Edy's Mills, s.e., under railroad ties, 4 May 72, JWR, 0-0-1. *Hwy 7, 1.61 km n of Arkona, under log, 4 May 72, JWR, 0-4-2. LANARK CO. *Hwy 43, 12.1 km e of Smiths Falls, under logs, 11 May 72, JWR, 4-0-6. *Hwy 43, 3.71 km e of Smiths Falls, under telephone pole in ditch, 11 May 72, JWR, 0-0-1. *Hwy 43, 5.32 km w of Smiths Falls, under logs, 11 May 72, JWR, 4-0-17. LEEDS CO. *Hwy 15, 2.26 km n of Seeley's Bay, under logs and concrete blocks in ditch, 16 May 72, JWR, 4-1-9. *Hwy 15, .97 km s of Portland, under logs, 16 May 72, JWR, 0-0-5. LENNOX AND ADDINGTON CO. *Hwy 2, Napanee, w.e., under rock behind EEEE Motel, 15 May 72, JWR, 0-0-1. *Hwy 2, 6.61 km w of Hwy 133, under log, 16 May 72, JWR, 0-0-1. MANITOULIN DIST. *Hwy 68, 5.97 km n of Birch Island, under rock, 13 May 72, JWR & JEM, 0-0-1. *Hwy 68, Birch Island, s.e., under logs, 13 May 72, JWR & JEM, 4-2-8. MIDDLESEX CO. Judd (1964). Reynolds (1972a). *Hwy 73, .97 km n of Mossley, under logs, 4 May 72, JWR, 0-0-2. *Hwy 7, 2.26 km s of Parkhill, under fence posts, 4 May 72, JWR, 0-1-6. NIAGARA CO. *Hwy 20, 1.61 km e of Smithville, old house site, 1 May 72, JWR, 0-1-4-1. NIPISSING DIST. *Hwy 17, 1.29 km e of Verner, under paper in wet ditch, 13 May 72, JWR & JEM, 0-0-3. NORTHUMBERLAND CO. *Hwy 45, 10.65 km s of Norwood, under logs next to barn, 28 Apr 72, JWR, 3-3-3-1. *Hwy 45, 4.84 km n of Baltimore, under log, 28 Apr 72, JWR, 0-0-1. ONTARIO CO. *Hwy 7, 1.61 km e of Green River, under logs, 26 Apr 72, JWR, 3-0-1. *Hwy 7-12, .81 km n of Myrtle, under logs, 26 Apr 72, JWR, 9-0-2. *Hwy 7, 5.32 km ne of Sunderland, under logs, 9 May 72, JWR, 0-0-4. OXFORD CO. *Hwy 59, 1.77 km e of Burgessville, under logs and wood chips, 1 May 72, JWR, 3-0-7. *Hwy 59, 1.13 km s of Curries, under junk, 1 May 72, JWR, 5-0-1. *Hwy 59, .32 km n of Hickson, under logs, 3 May 72, JWR, 0-0-2. PEEL CO. *Hwy 5, .81 km e of Dixie Rd, under logs, 29 Apr 72, JWR, 0-0-2. *Hwy 5, 4.35 km w of Hwy 10, under mattress, 29 Apr 72, JWR, 0-0-1. *Hwy 24, 8.87 km n of Erin, under log and dung, 29 Apr 72, JWR, 1-1-0. *Hwy 7, Brampton, e.e., under debris, 4 May 73, JWR, 0-1-4. PERTH CO. .81 km n of Fullarton, under rocks, 29 Apr 72, DWR & LWR, 2-1-3. *Mitchell, next to Collegiate, under logs, 4 May 72, JWR & DWR, 0-0-2. PETERBOROUGH CO. *Hwy 28, Lakefield College, under logs near waterfront, 27 Apr 72, JWR & CWR, 14-5-10. *Hwy 28, 5.48 km n of Burleigh Falls, under logs, 27 Apr 72, JWR, 1-0-6. PRESCOTT CO. *Hwy 34, 5.81 km n of Vankleek Hill, under logs, 11 May 72, JWR, 1-1-3. PRINCE EDWARD CO. *Hwy 33, 2.58 km e of Hillier, digging, 15 May 72, JWR, 1-0-0. *Hwy 49, Picton, n.e., under leaf pile, 15 May 72, JWR, 5-1-4. RUSSELL CO. *Hwy 17, 4.35 km w of Rockland, under log, 11 May 72, JWR, 2-0-1. SIMCOE CO. *Hwy 89, 5.48 km e of Alliston, under log, 2 May 72, JWR, 0-1-0. *Hwy 27, 5.97 km s of Cookstown, under logs, 2 May 72, JWR, 1-0-1. *Barrie, park opposite 14 Greenfield, under rocks and grass clippings, 7 May 72, JWR & GWA, 5-1-3. STORMONT CO. *Hwy 43, 1.29 km w of Finch, under logs, 11 May 72, JWR, 4-2-8. Hwy 43, 1.61 km e of Finch, under log, 11 May 72, JWR, 3-1-1. Hwy 43, 3.71 km w of Monkland, under logs, 11 May 72, JWR, 2-1-2. THUNDER BAY DIST. Hwy 17, 15.17 km e of Thunder Bay, small stream, 17 Jun 71, ROM Field Party, 9-0-0,

ROM-134. VICTORIA CO. *Hwy 46, 3.71 km n of Hwy 7, under log, 9 May 72, JWR, 0-0-1. *Hwy 46, .81 km n of Argyle, under logs, 9 May 72, JWR, 1-0-2. WATERLOO CO. *Hwy 97, 3.71 km n of Hwy 401, under log, 3 May 72, JWR, 0-0-1. Hwy 7-8, 1.77 km w of Petersburg, under log, 3 May 72, JWR, 1-2-0. Hwy 86, West Montrose, under log, 3 May 72, JWR, 0-0-1. Waterloo, Laurel Creek Conservation Area, 3 Aug 74, DPS, 0-0-1, UW-0007. WELLINGTON CO. *Hwy 24, 5.16 km e of Eramosa, under logs in cedar (*Thuja occidentalis*) woodlot, 29 Apr 72, JWR, 3-0-1. *Hwy 24, 4.03 km n of Erin, under logs, 29 Apr 72, JWR, 3-0-1. WENTWORTH CO. Reynolds (1972a). YORK CO. *Hwy Jct 48 and 401, Progress and Bellamy Rds, nw corner under lumber, 26 Apr 72, JWR, 1-1-7. *Hwy 48, .32 km n of Steeles Avenue, under logs, 26 Apr 72, JWR, 3-5-7. *Hwy 27, 1.45 km n of Hwy 7, under logs, 29 Apr 72, JWR, 1-4-9. *Edenbrook Park, Islington, under logs and rocks near stream bank, 30 Apr 72, JWR & DWR, 25-2-13. Edenbrook Park, Islington, quantitative study 1, formalin, 18 May 72, JWR, 16-3-13. Edenbrook Park, Islington, quantitative study 2, formalin, 18 May 72, JWR, 36-5-12-2.

Genus *Eiseniella* Michaelsen, 1900

1900 *Eiseniella* Michaelsen, Das Tierreich, Oligochaeta 10: 471.

Type Species

Enterion tetraedrum Savigny, 1826.

Diagnosis

Calciferous sacs, in x, digitiform, opening posteriorly into the gut ventrally in region of insertion of 10/11. Oesophagous of nearly uniform width through xi–xiv, calciferous channels narrow, lamellae low and continued along the lateral walls of the sacs. Intestinal origin, in xv. Gizzard, in xvii, weak, 17/18 not fenestrated. Typhlosole, simply lamelliform. Extraoesophageal trunks, joining dorsal vessel in xii. Hearts, in vii–xi. Nephridial bladder, short, sausage-shaped. Nephropores, inconspicuous, behind xv alternating irregularly and with asymmetry between a level just above *B* and one above *D*. Setae, not closely paired behind the clitellum. Prostomium epilobic. Longitudinal musculature, pinnate. (after Gates, 1972c: 108).

Discussion

Michaelsen (1900b) proposed the new name *Eiseniella* for a genus erected by Eisen in 1873. Although designated only as a new name for *Allurus* Eisen, 1873, Michaelsen included a second genus of Eisen's (*Tetragonurus* Eisen, 1874) in his *Eiseniella*. Both of Eisen's generic names were preoccupied (i.e., already in use as generic names in other groups); *Allurus* Foerster, 1862 had been used as a genus of hymenopterous insects, while *Tetragonurus* Risso, 1810 had been employed as a genus of fish. The type species for both of Eisen's genera are synonyms of *Enterion tetraedrum* Savigny, 1826. Insufficient data are available concerning the somatic anatomy of species included at various times in the classical *Eiseniella* to determine if they can be congeneric with *Eiseniella tetraedra*.

***Eiseniella tetraedra* (Savigny, 1826)**

Square-tail worm Ver à queue carrée
(Fig. 26)

1826 *Enterion tetraedrum* Savigny, Mém. Acad. Sci. Inst. Fr. 5: 184.
1826 ? *Lumbricus quadrangularis* Risso, Hist. Nat. Eur. Merid. 4: 426.
1828 ? *Lumbricus amphisbaena* Dugès, Ann. Sci. Nat. 15: 289.
1837 *Lumbricus tetraedrus*–Dugès, Ann. Sci. Nat., ser. 2, 8: 17, 23.
1843 *Lumbricus agilis* Hoffmeister, Arch. Naturg. 9(1): 191.
1871 *Lumbricus tetraedrus*–Eisen + *L. t. luteus* + *L. t. obscurus* Eisen, Öfv. Vet.-Akad. Förh. Stockholm 27(10): 966, 967, 968.
1873 *Allurus tetraedrus*–Eisen, Öfv. Vet.-Akad. Förh. Stockholm 30(8): 54.
1874 *Tetragonurus pupa* Eisen, Öfv. Vet.-Akad. Förh. Stockholm 31(2): 47.
1885 *Allurus neapolitanus* Örley, Ertek. Term. Magyar Akad. 15(18): 12.
1886 *Allurus ninnii* Rosa, Atti Ist. Veneto, ser. 6, 4: 680.
1889 *Lumbricus (Allolobophora) neapolitanus* + *L. (Allurus) tetraedrus* + *L. (Eisenia) pupa*–L. Vaillant, Hist. Nat. Annel. 3(1): 113, 151, 154.
1889 *Allurus hercynius*–Michaelsen + *A. dubius* Michaelsen + *A. ninnii*–Michaelsen, Mitt. Mus. Hamburg 7(3): 7, 10.
1890 *Eisenia pupa*–Benham, Quart. J. Micros. Soc., n.s., 31(2): 266.
1892 *Allolobophora tetragonurus*–Friend, Sci. Gossip 28: 194.
1892 *Allurus tetraedrus* + *A. amphisbaena* + *A. flavus* + *A. tetragonurus*–Friend, Proc. R. Irish Acad., ser. 3, 2: 402.
1896 *Allurus tetraedrus*–Ribaucourt + *A. bernensis* + *A. novis* + *A. infinitesimalis* Ribaucourt, Rev. Suisse Zool. 4: 69, 73, 74.
1900 *Eiseniella tetraedra*–Michaelsen, Das Tierreich, Oligochaeta 10: 471.
1937 *Eiseniella tetraedra* f. *typica*–Černosvitov, Rec. Indian Mus. 39: 107.
1974 *Eisenia tetraedra* (laps.)–Vail, Bull. Tall Timbers Res. Stn. 11: 2.

Diagnosis

Length 30–60 mm, diameter 2–4 mm, segment number 60–90, prostomium epilobic, first dorsal pore 4/5–5/6. Clitellum xxii, xxiii–xxvi, xxvii. Tubercula pubertatis uniformly broad on xxiii–xxv, xxvi. Setae closely paired, *AA:AB:BC:CD:DD* = 3:1:3:1:6–8 posteriorly. Ventral setae on x, or ix and x modified into genital setae. Male pores on xiii with slightly elevated glandular papillae in Černosvitov's "typical form". This is a misnomer because the normal or "typical" position for other members of the family is on xv. A second form of the same species, not recorded from Ontario, has male pores on xv. Seminal vesicles, four pairs on 9–12. Spermathecae, two pairs opening between *d* and mD line in 9/10 and10/11. Body cylindrical in front of clitellum and quadrangular behind. Colour variable, from dark brown, greenish, ruddy to bright golden yellow.

Biology

Eiseniella tetraedra is a limicolous species and shows a marked preference for damp habitats. It is known from wells, springs, subterranean waters, rivers,

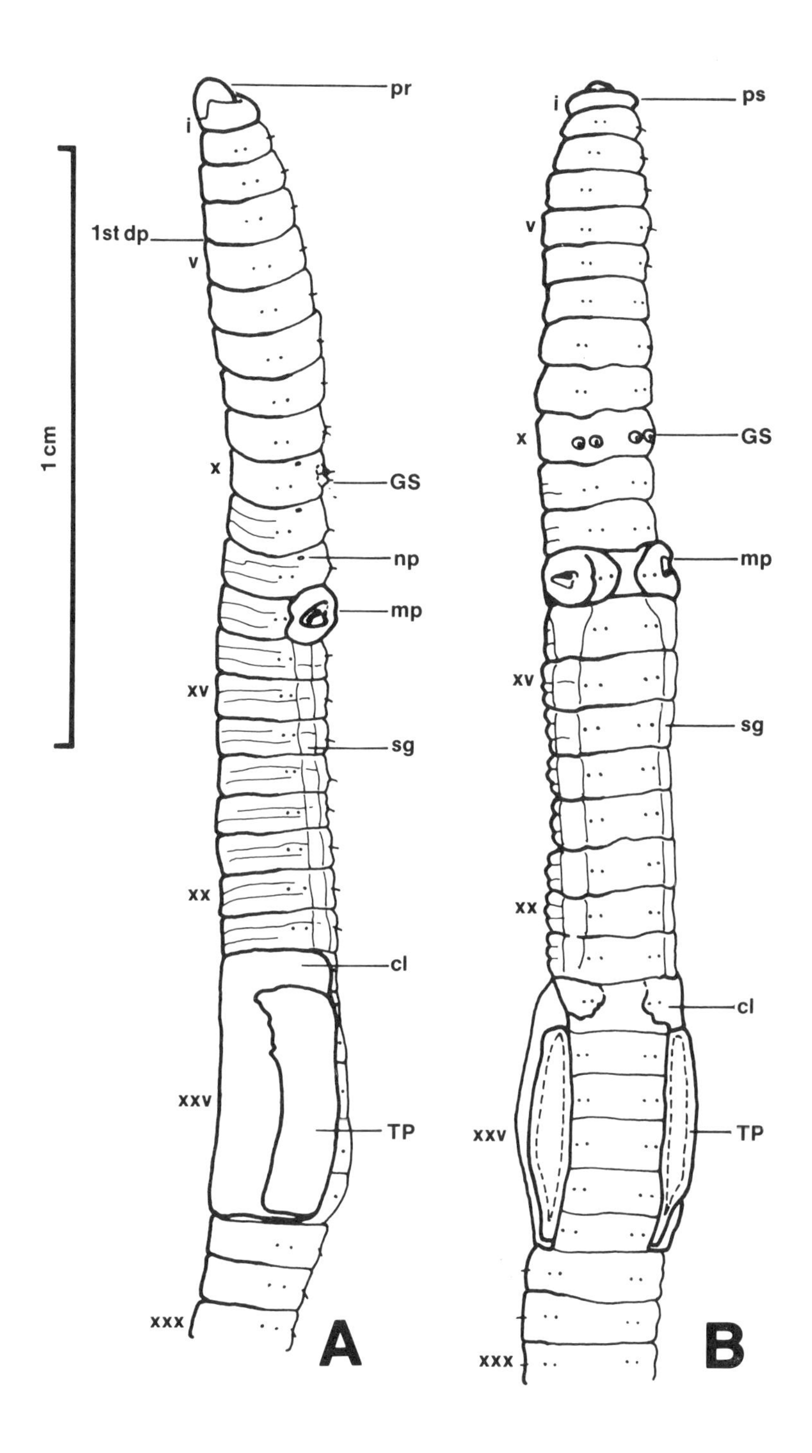

Fig. 26 External longitudinal views of *Eiseniella tetraedra* showing taxonomic characters. A. Lateral view. B. Ventral view. (ONT: Huron Co., cat. no. 7549)

ponds, lakes, and canals, and may be one of the dominant animals in the dense moss of swift streams (Gates, 1972c). In Ohio it has been recorded from soils with a pH range of 6.8 to 8.5, a moisture content of 25–35%, and an organic matter content of 4–5%. It is the most common megadrile in British caves and is known from caves in the rest of Europe and in South Africa. Olson (1928, 1936) and Eaton (1942) reported this species from water-soaked banks of streams, lakes, and ponds. Murchie (1956) reported it from the bottom deposits of streams, lakes, or ponds, from wet to very moist stream banks and lake shores, from bottom lands subject to flooding, or with a high water table, and from seepage areas around springs at upland sites. The soil type for these sites varied from peaty organic material to sandy gravel. The sources of specimens for the present study were all moist to wet habitats.

Under favourable conditions activity can be year round but in Ontario there are probably summer and winter rest periods. Aestivation involves immobility and tight coiling in a small mucus-lined cavity; whether hibernation involves quiescence or diapause is not known. This species is obligatorily parthenogenetic (Muldal, 1952; Omodeo, 1955b; Reynolds, 1974c). The first reports of uniparental reproduction for megadriles involved experiments with *Eiseniella tetraedra* (Gavrilov, 1935, 1939).

Range

A native of Palaearctis, *El. tetraedra* is now known from Europe, North America, South America, Asia, Africa, and Australasia (Gates, 1972). It also occurs in Iceland (Backlund, 1949). It is another cosmopolitan species that has been carried around the world.

North American Distribution

British Columbia (Gates, 1972c), New Brunswick (Reynolds, 1976d), Nova Scotia (Reynolds, 1975a, 1976a), Ontario (Eisen, 1874), Prince Edward Island (Reynolds, 1975c), Québec (Reynolds, 1975b, d, e, 1976c), Arizona (Reynolds et al., 1974), Arkansas (Causey, 1952), California (Smith 1917), Connecticut (Reynolds, 1973c), District of Columbia (Eaton, 1942), Idaho (Gates, 1967), Illinois (Smith, 1917), Indiana (Heimburger, 1915), Louisiana (Harman, 1952), Maine (Gates, 1966), Maryland (Reynolds, 1974b), Massachusetts (Reynolds, 1977), Michigan (Smith 1917), Missouri (Olson, 1936), Montana (Reynolds, 1972c), Nevada (Gates, 1967), New Hampshire (Reynolds, 1976c), New Jersey (Davies, 1954), New York (Olson, 1940), North Carolina (Pearse, 1946), Ohio (Olson, 1928), Oregon (MacNab and McKey-Fender, 1947), Pennsylvania (Moore, H.F., 1895), Rhode Island (Reynolds, 1973b), South Dakota (Gates, 1967), Tennessee (Reynolds, 1974a), Utah (Gates, 1967), Vermont (Reynolds, 1976c), Virginia (Harman, 1960), Washington (Smith, 1917), West Virginia (Reynolds, 1974b), Wisconsin (Gates, 1972c), Wyoming (Gates, 1967). New records: Alaska, Georgia, Iowa, Kentucky.

Ontario Distribution (Fig. 27)

Eiseniella tetraedra was the fourth species reported from Ontario by Eisen (1874). This species has also been recorded in the province by Stafford (1902) and Judd (1964). The distribution of *El. tetraedra* in Ontario is generally close to the Great Lakes and concentrated in southwestern Ontario.

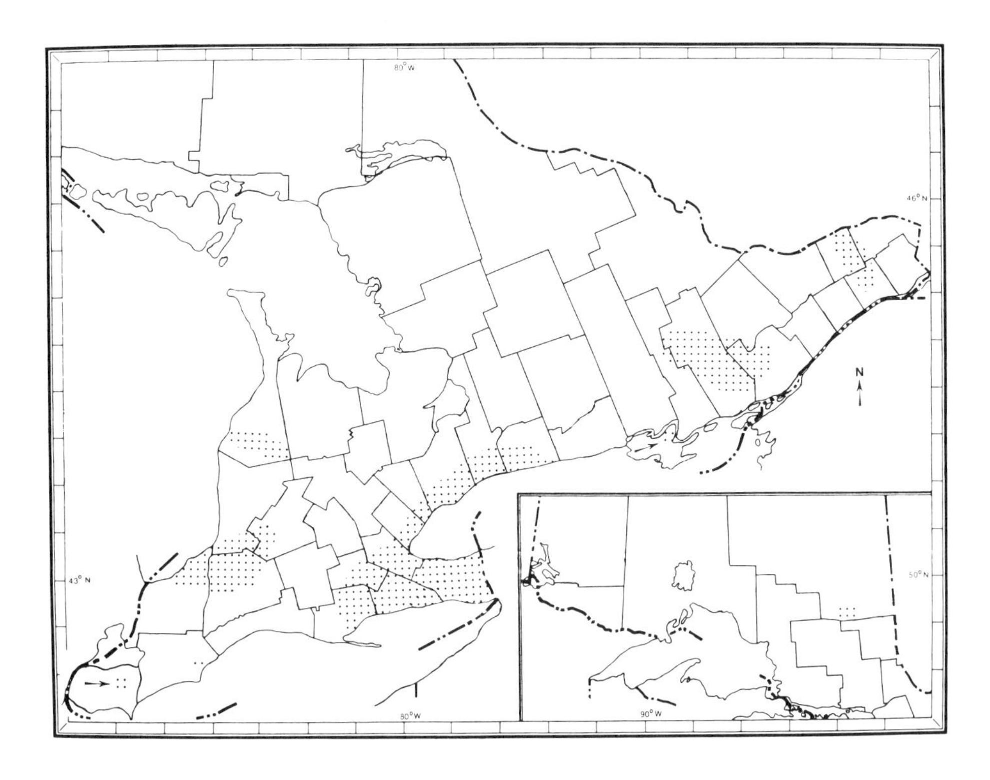

Fig. 27 The known Ontario distribution of *Eiseniella tetraedra*.

BRUCE CO. *Hwy 4, 6.77 km s of Teeswater, under log, 5 May 72, JWR, 0-1-0. *Hwy 21, 6.94 km n of Kincardine, under log, 5 May 72, JWR, 0-0-1. COCHRANE DIST. Reynolds (1972a). DURHAM CO. Kendal, cold fast stream, 31 May 66, IMS, 4-4-7, ROM-I1. ESSEX CO. *Hwy 3, Ruthven, n.e., wet ditch, next to railroad tracks, 4 May 72, JWR, 0-0-2. FRONTENAC CO. *Hwy 15, 4.03 km n of Hwy 401, under paper in wet ditch, 16 May 72, JWR, 1-0-1. *Hwy 41, 1.29 km n of Cloyne, under log in ditch, 16 May 72, JWR, 0-0-1. HALDIMAND CO. *Hwy 3, 6.13 km w of Cayuga, under logs, 1 May 72, JWR, 0-2-1. HURON CO. *Hwy 4, 1.61 km n of Exeter, wet ditch, 5 May 72, JWR, 1-1-12. KENT CO. Morpeth, under stones near creek, 21 Apr 75, DRB, 0-0-1, UW-0001. LAMBTON CO. Hwy 21, 2.1 km n of Wyoming, under logs in wet ditch, 4 May 72, JWR, 0-0-3. LEEDS CO. *Hwy 15, 2.26 km n of Seeley's Bay, under logs and concrete blocks in ditch, 16 May 72, JWR, 0-0-3. Chaffey's Locks, swamp near Lake Opinicon, 4 Sep 65, IMS, 0-0-1, ROM-II7. MIDDLESEX CO. Judd (1964). NIAGARA CO. Eisen (1874). NIPISSING DIST. *Hwy 17, 1.29 km e of Verner, under paper in wet ditch, 13 May 72, JWR & JEM, 0-1-1. NORFOLK CO. Reynolds (1972a), NMC-1076. *Hwy 6, 4.84 km s of Jarvis, under logs, 1 May 72, JWR, 0-2-0. ONTARIO CO. Dufferin Creek area, day after use of lampricide, 11 May 71, TY, 0-0-5, ROM-I38. OXFORD CO. Reynolds (1972a), NMC-1077. PEEL CO. Port Credit, Credit River, 6 May 66, IMS, 1-0-0, ROM. PERTH CO. Hwy Jct 23 and 83, Russeldale, under paper in wet ditch, 5 May 72, JWR, 4-0-3. PRINCE EDWARD CO. *Hwy 49, 14.84 km n of Picton, under paper in ditch, 15 May 72, JWR, 2-1-13. RUSSELL CO. *Hwy 17, 7.74 km e of Rockland, under paper in wet ditch, 11 May 72, JWR, 1-0-2. STORMONT CO. *Hwy 43, 5.48 km w of Finch, in and under straw in wet ditch, 11 May 72, JWR, 0-0-3. WENTWORTH CO. Reynolds (1972a). Strathcona Gardens, Lake Ontario, 7 May 75, DRB, 0-0-2, UW-0001. YORK CO. *Edenbrook Park, Islington, under log near stream bank, 30 Apr 72, JWR & DWR, 0-0-1.

Genus *Lumbricus* Linnaeus, 1758

1758 *Lumbricus* (part.) Linnaeus, Syst. Nat. (ed. 10), p. 647.
1774 *Lumbricus* (part.)–Müller, Verm. Terr. Fluv. 1(2): 24.
1780 *Lumbricus* (part.)–Fabricius, Fauna Grønlandica, p. 277.
1826 *Enterion* (part.) Savigny, Mém. Acad. Sci. Inst. Fr. 5: 179.
1836 *Lumbricus* (part.)–Templeton + *Omilurus* Templeton, Ann. Mag. Nat. Hist. 9: 235.
1845 *Lumbricus* (part.)–Hoffmeister, Regenwürmer, p. 4.
1873 *Lumbricus*–Eisen, Öfv. Vet.-Akad. Förh. Stockholm 30(8): 45.
1876 *Lumbricus* (part.)–Claus, Grundzüge der Zool. (ed. 3) 1: 416.
1880 *Lumbricus* (part.)–Claus, Grundzüge der Zool. (ed. 4) 1: 478.
1881 *Lumbricus* (part.) + *Enterion*–Örley, Math. Term. Közlem. Magyar Akad. 16: 580, 587.
1894 *Allolobophora* (part.)–W.W. Smith, Trans. N.Z. Inst. 26: 117.
1900 *Lumbricus*–Michaelsen, Das Tierreich, Oligochaeta 10: 508.
1930 *Lumbricus*–Stephenson, Oligochaeta, p. 914.
1975 *Lumbricus*–Gates, Megadrilogica 2(1): 3.

Type Species

Lumbricus terrestris Linnaeus, 1758, by Sims (1973).

Diagnosis

Calciferous sacs, in x, digitiform to pyriform, opening into gut posteriorly and ventrally in region of insertion of 10/11, lamellae continued along lateral walls. Oesophagus, widened and markedly moniliform in xi–xii, in those segments with a vertically slitlike lumen which widens as lamellae gradually narrow behind 12/13. Intestinal origin, in xv. Gizzard, mainly in xvii. Typhlosole, high, rather thick and nearly oblong vertically, grooves not continuous across ventral face. Extraoesophageal trunks, joining dorsal trunk in region of ix–x. Hearts, in vii–xi. Nephridial bladder, *J*-shaped, closed end laterally, duct passing into parietes near *B*. Nephropores, obvious, behind xv irregularly alternating, with asymmetry, between levels just above *B* and well above *D*. Setae, closely paired. First dorsal pore, anterior to 10/11. Prostomium, tanylobic. Body compressed dorsoventrally behind clitellum and with a more or less trapezoidal transverse section. Longitudinal musculature, pinnate. Pigment, lacking immediately underneath intersegmental furrows, in circular muscle layer (after Gates, 1972c: 113, 1975a: 3).

Discussion

The genus *Lumbricus* (Linnaeus, 1758, p. 647) originally contained only two species, *L. terrestris* and *L. marinus*. Since the latter was not a member of the Oligochaeta, the type species was declared to be *Lumbricus terrestris* (I.C.Z.N., Opinion 75). Discussions of what was really meant by *L. terrestris* Linnaeus, 1758 (and his lengthened definition of Syst. Nat., ed. 12, 1767) have raged ever since but were officially settled recently (Sims, 1973; Gates, 1973b; and Bouché, 1973). The result of this action was the neotypification of *Lumbricus terrestris* Linnaeus by Sims (1973) with an expanded definition, and the deposition of type material from Sweden in the British Museum (Natural History).

Lumbricus castaneus (Savigny, 1826)

Chestnut worm Ver alezan

(Fig. 28)

1826 *Enterion castaneus* + *E. pumilum* Savigny, Mém. Acad. Sci. Inst. Fr. 5: 180, 181.

1837 *Lumbricus castaneus*–Dugès, Ann. Sci. Nat., ser. 2, 8: 17, 22.

1851 *Lumbricus triannularis* Grube, *In:* Middendorff, Reise Sibirien 2(1): 18.

1865 *Lumbricus minor* Johnston, Cat. British Non-paras. worms, p. 59.

1867 *Lumbricus josephinae* Kinberg, Öfv. Vet.-Akad. Förh. Stockholm 23: 98.

1871 *Lumbricus purpureus* Eisen, Öfv. Vet.-Akad. Förh. Stockholm 27(10): 956.

1881 *Enterion pupureum* + *Lumbricus purpureus*–Örley, Math. Term. Közlem. Magyar Akad. 16: 588, 590.

1889 *Lumbricus (L.) castaneus* + *L. (L.) purpureus* + *L. (L.) triannularis*–L. Vaillant, Hist. Nat. Annel. 3(1): 124, 127, 129.

1894 *Allolobophora purpureus*–W.W. Smith, Trans. N.Z. Inst. 26: 117.

1895 *Lumbricus pumilosum* (laps.)–Beddard, Monogr. Oligo. (Oxford), p. 722.

1896 *Lumbricus castaneus*–Ribaucourt + *L. morelli* Ribaucourt + *L. perrieri* Ribaucourt, Rev. Suisse Zool. 4: 10, 13, 14.

1900 *Lumbricus castaneus*–Michaelsen, Das Tierreich, Oligochaeta 10: 510.

1936 *Lumbricus castaneus* var. *disjonctus* Tétry, Bull. Soc. Sci. Nancy, n. ser. 1936: 196.

1957 *Lumbricus castaneus* var. *pictus* Chandebois, Bull. Soc. Zool. France 82: 417.

1972 *Lumbricus castaneus*–Bouché, Inst. Natn. Rech. Agron., p. 362.

Diagnosis

Length, 30–50 mm (generally $<$35mm), diameter 3–5 mm, segment number 70–100, prostomium tanylobic, first dorsal pore 5/6–8/9. Clitellum xxviii–xxxiii. Tubercula pubertatis xxix–xxxii. Setae closely paired, $AA \simeq BC$, $AB > CD$, $DD \simeq$ ½C anteriorly and $DD <$ ½C posteriorly. Setae *a* and *b* on ix and/or x on pale genital tumescences fused ventrally. Male pores inconspicuous on xv. Seminal vesicles, three pairs in 9, 11, and 12+13. Spermathecae, two pairs with short ducts in 9/10 and 10/11. Colour, deeply pigmented, dark red, chestnut, violet brown and strongly iridescent. Body cylindrical and dorsoventrally flattened posteriorly.

Biology

Lumbricus castaneus has been recorded from soils with a pH range of 4.6–8.0, from gardens, cultivated fields, pastures, forests, taiga, steppes, among organic matter such as manure and compost or leaf litter, and in banks by water (Gates, 1972c). It has been found in caves in Europe. Gerard (1964) reported it as terrestrial, mostly in soil rich in organic matter, in gardens, parks, pastures, forests, on river and marsh banks, and under stones, leaves and dung. Apart from this author's records all North American records fail to give habitat information but my findings (Reynolds, 1973a, 1973b, 1973c, 1974b, 1975a-e) are similar to those of Gerard in England. In Ontario *L. castaneus* was collected from a stream

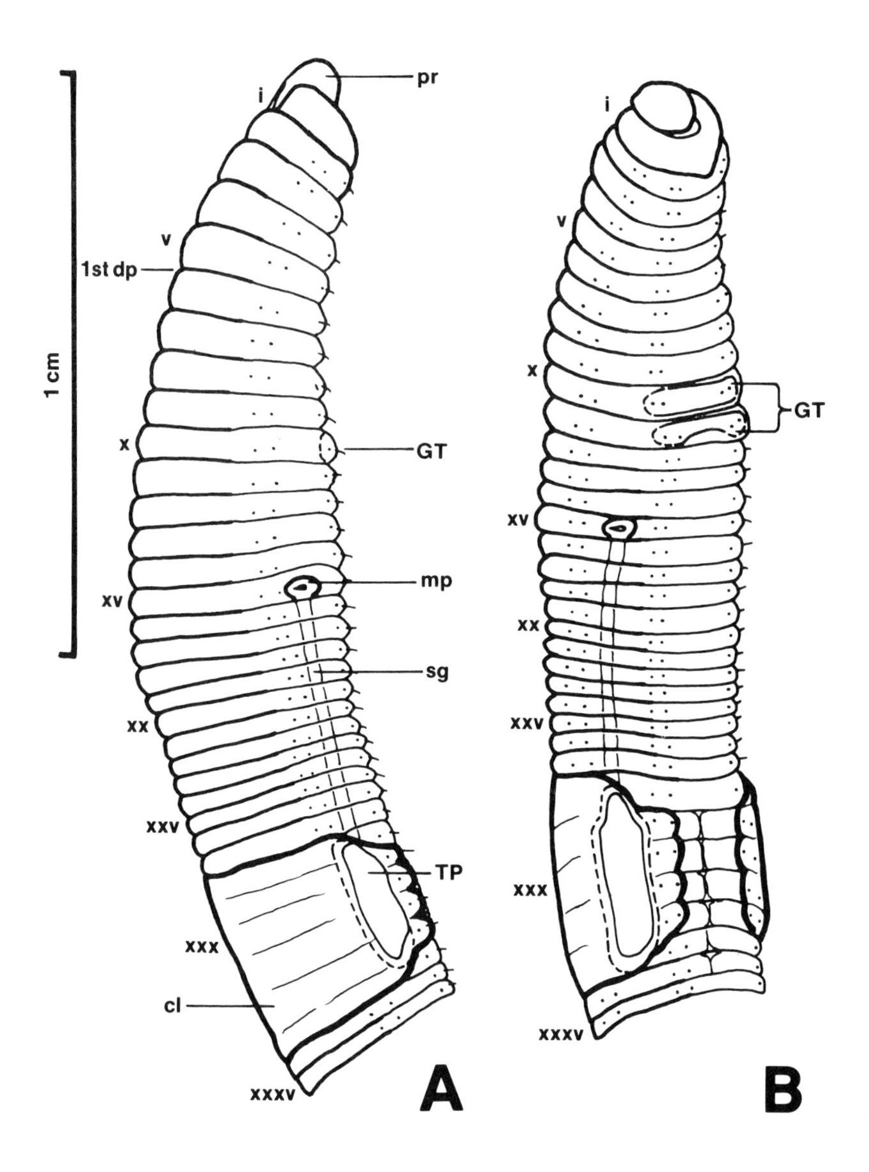

Fig. 28 External longitudinal views of *Lumbricus castaneus* showing taxonomic characters. A. Lateral view. (NS: Queens Co., cat. no. 8880) B. Ventrolateral view. (NS: Digby Co., cat. no. 8889)

bank (Reynolds, 1972a) and Judd (1964) obtained his specimens from the Byron Bog near London.

In unfavourable seasons individuals may be found 1 to 1½ metres down in the soil, but little is known of the annual cycle. Copulation, which may not involve secretion of a slime tube (cf. pp. 3-4), is subterranean. This species is obligatorily amphimictic (Reynolds, 1974c). Janda and Gavrilov (1939) did not find that isolated individuals laid cocoons.

Range

A native of Palaearctis, *L. castaneus* is now known from Europe, North America including Mexico, New Zealand, and St. Helena (Gates, 1972c). It also occurs in Iceland (Backlund, 1949).

North American Distribution

New Brunswick (Reynolds 1976d), Newfoundland (Gates, 1942), Nova Scotia (Reynolds, 1975a, 1976a), Ontario (Eisen, 1874), Prince Edward Island (Reynolds, 1975c), Québec (Reynolds, 1975b, d, e, 1976c), Delaware (Reynolds, 1973a), Connecticut (Reynolds, 1973c), Idaho (Gates, 1967), Maine (Gates, 1966), Maryland (Reynolds, 1974b), Massachusetts (Reynolds, 1977), New York (Smith, 1917), Oregon (Gates, 1972c), Pennsylvania (Bhatti, 1965), Rhode Island (Reynolds, 1973b), Vermont (Reynolds, 1976c), Virginia (Reynolds, 1974b), West Virginia (Reynolds, 1974b). New records: Alaska, Washington.

Ontario Distribution (Fig. 29)

Lumbricus castaneus was the fifth species reported from Ontario by Eisen (1874). It has been recorded only twice since (Judd, 1964 and Reynolds, 1972a). This species has been recorded from only four counties in the province.

HALIBURTON CO. *Reynolds (1972a). MIDDLESEX CO. Judd (1964). NIAGARA CO. Eisen (1874). YORK CO. Scarborough Bluffs, Brimley Rd, garden, 28 Oct 41, JO, 0-0-1, ROM.

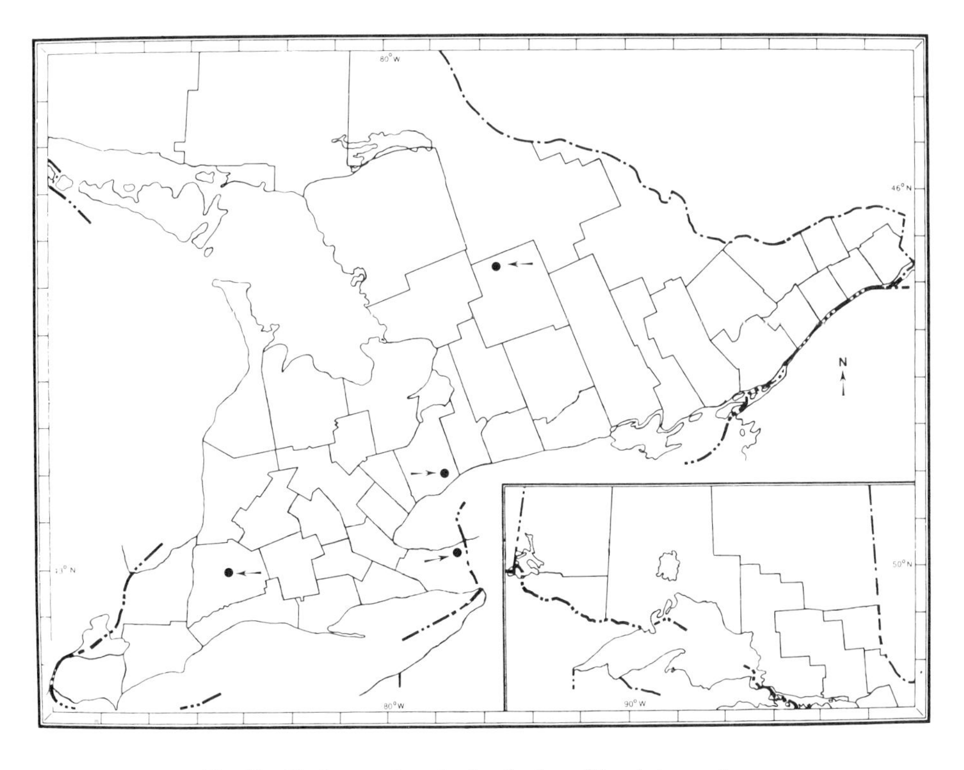

Fig. 29 The known Ontario distribution of *Lumbricus castaneus*.

Lumbricus festivus (Savigny, 1826)

Quebec worm Ver québécois

(Fig. 30)

1826 *Enterion festivum* Savigny, Mém. Acad. Sci. Inst. Fr. 5: 180.
1836 *Lumbricus omilurus* R. Templeton, Ann. Mag. Nat. Hist. 9: 235.
1837 *Lumbricus festivus*–Dugès, Ann. Sci. Nat., ser. 2, 8: 17, 21.
1891 *Lumbricus rubescens* Friend, Nature 44: 273.
1900 *Lumbricus festivus*–Michaelsen, Das Tierreich, Oligochaeta 10: 512.

Diagnosis

Length 48–105 mm, diameter 4–5 mm, segment number 100–143, prostomium tanylobic, first dorsal pore 5/6. Clitellum xxxiv–xxxix. Tubercula pubertatis xxxv–xxxvii, xxxviii. Setae closely paired, $AA:AB:BC:CD$ = 34:12:25:8 anteriorly, and 35:10:25:8 posteriorly. Setae on segments v–x are notably enlarged and more widely paired. Some of the ventral setae on viii–xiv, xviii, xxv–xxxix are on genital tumescences. Male pores on xv with glandular papillae extending onto xiv and xvi. Seminal vesicles, three pairs in 9, 11, and 12+13+14. Spermathecae, two pairs with short ducts opening on level *c* in 9/10 and 10/11. Colour, ruddy brown, iridescent dorsally and lightly coloured ventrally. Body cylindrical and slightly dorsoventrally flattened posteriorly.

Biology

Lumbricus festivus is rare and little is known about its habits. Gerard (1964) recorded the habitats of this species as pastures, on river banks, in soil beneath dung, leaves, and under stones. I found the species at ten sites in Ontario, six times under logs, three times under debris, and once under a leaf pile.

Lumbricus festivus is assumed to be obligatorily amphimictic (Reynolds, 1974c) and copulation occurs beneath the soil surface. In recent surveys (Reynolds, 1975d, 1975e, 1976c) specimens with spermatophores have been found.

Range

Known only from western Europe and northeastern North America, notably southern Québec.

North American Distribution

New Brunswick (Langmaid, 1964), Québec (Reynolds, 1975d, e, 1976c), Vermont (Reynolds, 1976c). New record: British Columbia.

Ontario Distribution (Fig. 31)

Lumbricus festivus was first reported from Ontario by Stafford (1902) but his report simply stated "from Ontario, Quebec, New Brunswick and Nova Scotia" with no habitat or locality information. I have been unable to verify the existence of this species in New Brunswick or Nova Scotia (Reynolds, 1975a, 1976a, 1976d). The specimens obtained during this study are the first reports of *L. festivus* in Ontario since 1902.

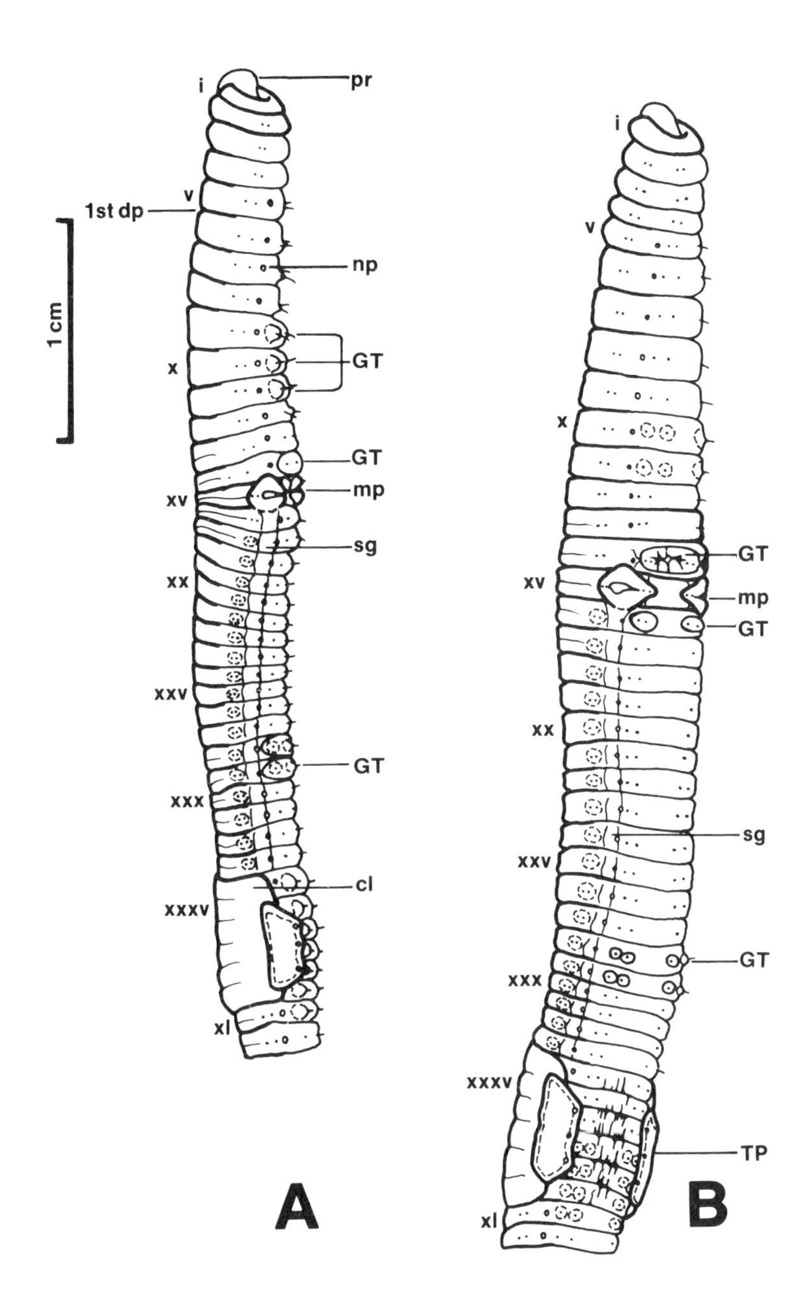

Fig. 30 External longitudinal views of *Lumbricus festivus* showing taxonomic characters. A. Lateral view. (ONT: Glengarry Co., cat. nos. 7620 and 7621) B. Ventrolateral view. (ONT: Northumberland Co., cat. no. 8108)

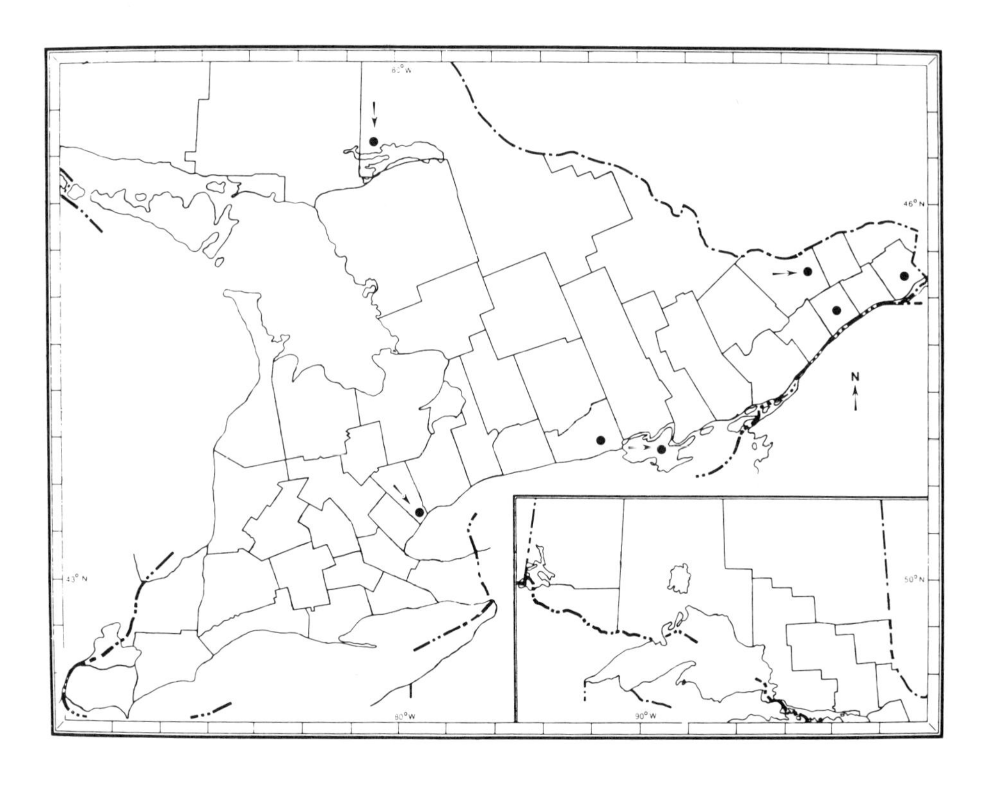

Fig. 31 The known Ontario distribution of *Lumbricus festivus*.

CARLETON CO. *Ottawa-Carleton Rd 35, 2.42 km s of Leonard, under logs, 11 May 72, JWR, 1-1-0. DUNDAS CO. *Hwy 31, 3.55 km n of Morrisburg, under lumber, 11 May 72, 0-1-0. GLENGARRY CO. *Hwy 34, 11.13 km n of Lancaster, under log, 11 May 72, 1-0-0. *Hwy 34, 5.65 km n of Alexandria, under log, 11 May 72, JWR, 0-0-1. NIPISSING DIST. *Hwy 17, 1.29 km e of Verner, under paper in wet ditch, 13 May 72, JWR & JEM, 2-1-0. NORTHUMBERLAND CO. *Hwy 45, Baltimore, under junk, 28 Apr 72, JWR, 1-4-1. *Hwy 401, 5.65 km e of Grafton, under logs, 15 May 72, JWR, 3-1-1. *Northumberland-Durham Rd 1, .65 km w of Hwy 33, under logs, 15 May 72, JWR, 1-1-1. PEEL CO. *Hwy 5, .81 km e of Dixie Rd, under log, 29 Apr 72, JWR, 0-1-0. PRINCE EDWARD CO. *Hwy 49, Picton, n.e., under leaf pile, 15 May 72, JWR, 1-0-0.

Lumbricus rubellus **Hoffmeister, 1843**

Red marsh worm Ver rouge du marécage

(Fig. 32)

1843 *Lumbricus rubellus* Hoffmeister, Arch. Naturg. 9(1): 187.
1877 *Lumbricus campestris* (part.) Hutton, Trans. N.Z. Inst. 9: 351.
1881 *Enterion rubellum* var. *parvum* + *E. r.* var. *magnum* Örley, Math. Term. Közlem. Magyar Akad. 16: 588, 589.
1883 *Digaster campestris* (part.)–Hutton, N.Z. J. Sci. 1: 586.

1887 *Endrilus campestris* (part.)–W.W. Smith, Trans. N.Z. Inst. 19: 137.
1892 *Lumbricus rubellus* var. *curticaudatus* Friend, J. Linn. Soc. Lond. 24: 312.
1894 *Allolobophora rubellus*–W.W. Smith, Trans. N.Z. Inst. 26: 117.
1900 *Lumbricus rubellus*–Michaelsen, Das Tierreich, Oligochaeta 10: 509.
1909 *Allolobophora relicta* Southern, Proc. R. Irish Acad. 27B(8): 119.
1923 *Lumbricus rubellus*–Stephenson, Fauna British India, Oligochaeta, p. 508.
1972 *Lumbricus rubellus rubellus*– Bouché + *L. r. castaneoides* + *L. r. friendoides* Bouché, Inst. Natn. Rech. Agron., p. 368, 371, 372.

Diagnosis

Length 50–150 mm (usually >60 mm), diameter 4–6 mm, segment number 70–120, prostomium tanylobic, first dorsal pore 5/6–8/9. Clitellum xxvi, xxvii–xxxi, xxxii. Tubercula pubertatis on xxviii–xxxi. Setae closely paired, *AA* > *BC*, *AB* > *CD*, *DD* = ½C posteriorly. Genital tumescences in viii–xii (less frequently on x), xx–xxiii, xxvi–xxxvi. Male pores, inconspicuous, without glandular papillae on xv. Seminal vesicles, three pairs in 9, 11, and 12+13. Spermathecae, two pairs with short ducts opening in 9/10 and 10/11. Colour, ruddy brown or red-violet and iridescent dorsally, pale yellow ventrally. Body cylindrical and sometimes dorsoventrally flattened posteriorly.

Biology

Lumbricus rubellus has been recorded from natural soils of pH 3.8–8.0 and shows a wide tolerance of habitat factors. Olson (1928, 1936) reported it from under debris. Eaton (1942) found it in stream banks, under logs, and in woody peat and stated that it seemed to require a great deal of moisture and organic matter. Černosvitov and Evans (1947) recorded the habitats of this species as places rich in humus, abundant in parks, gardens, pastures, on river banks, under stones, moss, or old leaves. Gerard (1964) also found this species frequently aggregated beneath dung in pastures as well as the sites mentioned above. In Ontario, *L. rubellus* was obtained from a wide variety of habitats. This species also is known from caves in Europe.

Under suitable conditions activity, including breeding, is year round. *L. rubellus* is obligatorily amphimictic (Reynolds, 1974c) and copulation, like defecation, occurs below the soil surface, or in the litter layer, at any time of day. It seems that copulation does not involve a mucous tube (Gates, 1972c).

This species has been cultured by the fish bait industry (Gates, 1972c). It is also important in the decomposition of litter.

Range

Lumbricus rubellus is now known from Europe, Iceland, North America, Mexico, Asia, South Africa, and New Zealand (Gates, 1972c).

North American Distribution

British Columbia (Berkeley, 1968), New Brunswick (Reynolds, 1976d), Newfoundland (Smith, 1917), Nova Scotia (Reynolds, 1975a, 1976a), Ontario (Reynolds, 1972a), Prince Edward Island (Reynolds, 1975c), Québec (Reynolds, 1975b, e, 1976c), Alaska (Gates, 1954), Arkansas (Causey,

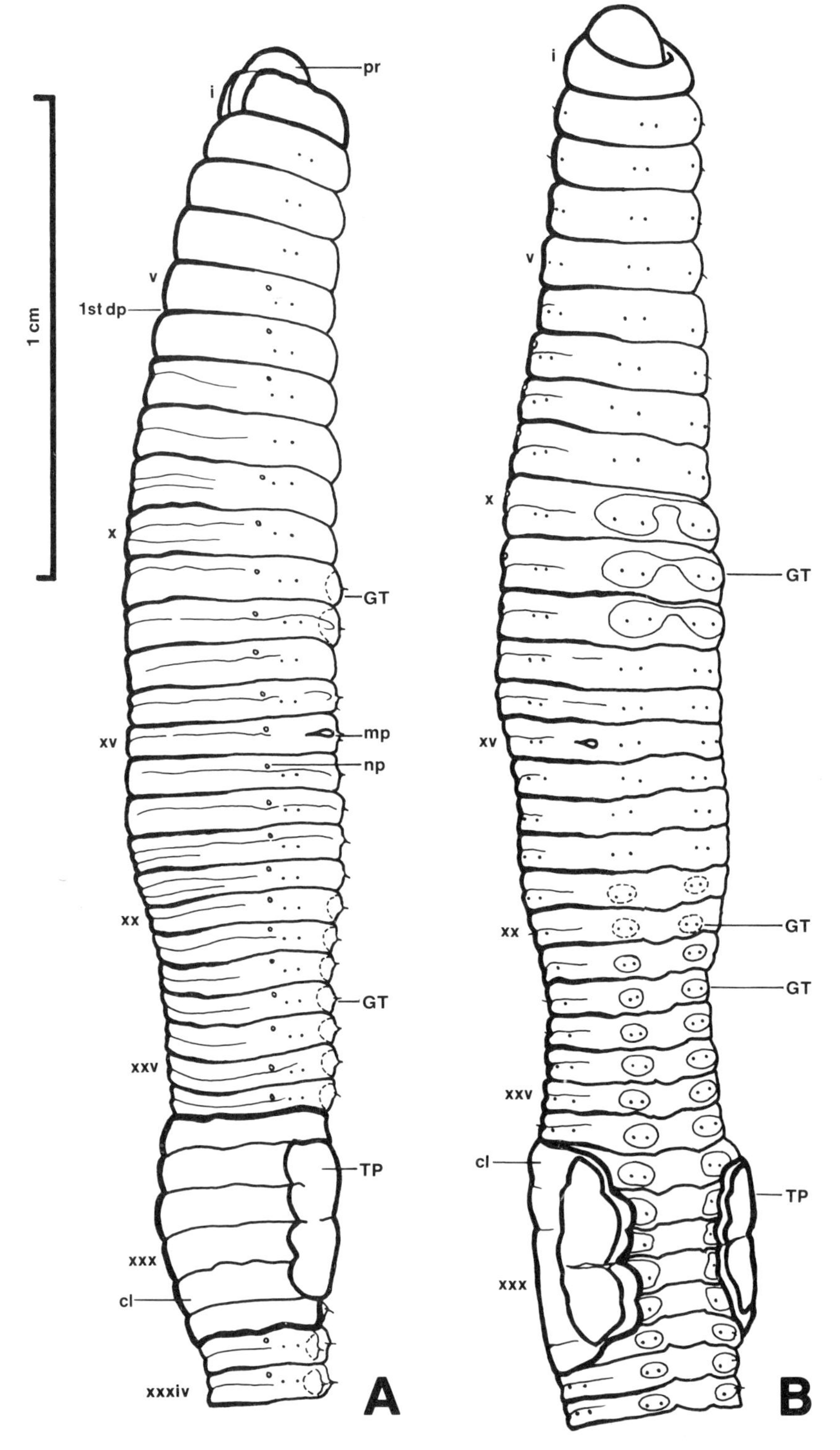

Fig. 32 External longitudinal views of *Lumbricus rubellus* showing taxonomic characters. A. Lateral view. B. Ventral view. (ONT: Parry Sound Dist., cat. no. 7505)

1952), California (Smith 1917), Colorado (Gates, 1967), Connecticut (Reynolds, 1973c), Delaware (Reynolds, 1973a), District of Columbia (Eaton, 1942), Florida (Gates, 1972c), Idaho (Gates, 1967), Illinois (Smith, 1928), Indiana (Joyner, 1960), Maryland (Reynolds, 1974b), Massachusetts (Reynolds, 1977), Michigan (Smith, 1917), Missouri (Olson, 1936), New Hampshire (Reynolds, 1976c), New Jersey (Davies, 1954), New York (Olson, 1940), Ohio (Olson, 1928), Oregon (Smith, 1917), Pennsylvania (Eaton, 1942), Rhode Island (Reynolds, 1973b), Tennessee (Reynolds, 1972b), Utah (Gates, 1967), Vermont (Reynolds, 1976c), Virginia (Reynolds, 1974b), Washington, (Smith, 1917), West Virginia (Reynolds, 1974b), Greenland (Omodeo, 1955a). New records: Manitoba, Georgia, Minnesota, North Carolina.

Ontario Distribution (Fig. 33)

Lumbricus rubellus was first reported from Ontario by Stafford (1902) and only once since then (Reynolds, 1972a). The species is widely distributed across the province and additional collecting under favourable conditions should fill the gaps in its distribution.

ALGOMA DIST. *Hwy 17, 7.58 km e of Spanish, under logs, 13 May 72, JWR & JEM, 16-10-8. *Hwy 17, .48 km e of Spanish, under logs and paper in ditch, 13 May 72, JWR & JEM, 4-1-5. Gargantua Bay, 11 Jun 75, DRB, 1-0-0, UW-0001. BRANT CO. *Hwy 53, 5.48 km w of Cathcart, under railroad ties, 1 May 72, JWR, 0-1-0. *Hwy 54, .64 km w of Onondaga, under log in wet area, 3 May 72, JWR, 0-1-0. BRUCE CO. Hwy 21, 2.57 km n of Tiverton, under lumber, 5 May 72, JWR, 1-0-0. CARLETON CO. Reynolds (1972a). COCHRANE DIST. Reynolds (1972a). DUFFERIN CO. *Hwy 9, 10–16 km w of Orangeville, under junk, 2 May 72, JWR, 0-1-0. *Hwy 9, 3.71 km w of Orangeville, under log in ditch, 2 May 72, JWR, 0-1-0. DUNDAS CO. *Hwy 34, 1.29 km e of Chester-

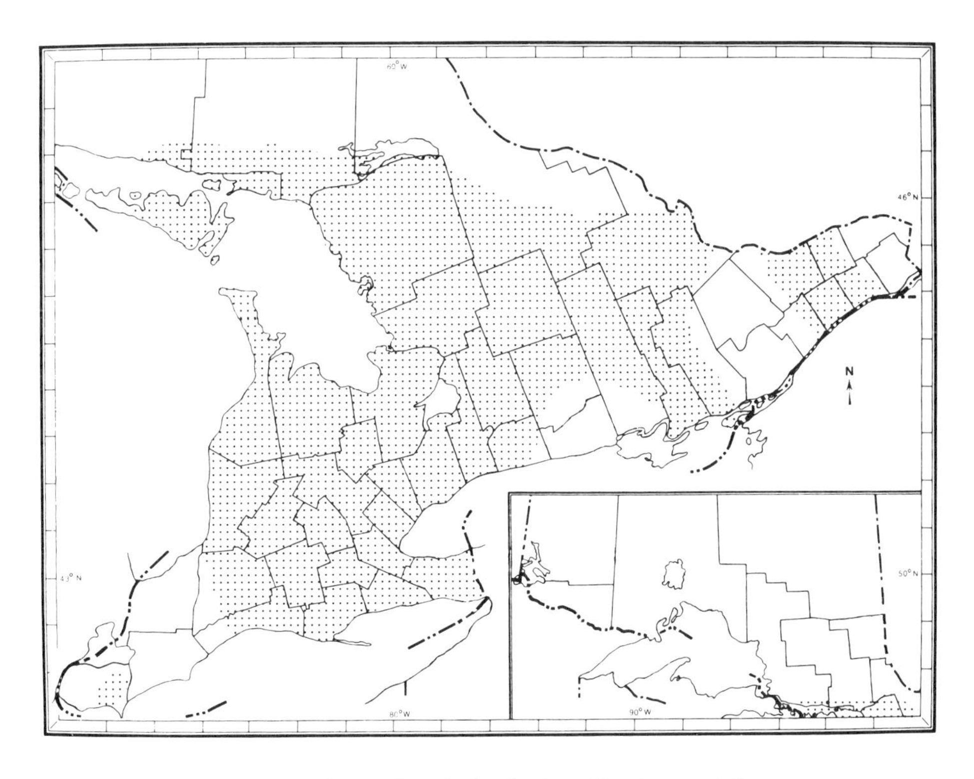

Fig. 33 The known Ontario distribution of *Lumbricus rubellus*.

ville, under rock in ditch, 11 May 72, JWR, 0-1-0. *Hwy 31, .81 km n of Hwy 43, under logs, 11 May 72, JWR, 1-1-4. DURHAM CO. *Hwy 401, .16 km w of Liberty Rd, Bowmanville, digging under log next to railroad tracks, 15 May 72, JWR, 1-0-0. *Hwy 401, .32 km e of Mill St. Newcastle, under log, 15 May 72, JWR, 1-0-0. ELGIN CO. *Hwy 73, 3.06 km s of Harrietsville, under rotten log, 4 May 72, JWR, 2-0-0. *Hwy 73, 6.77 km n of Aylmer, under logs, 4 May 72, JWR, 4-2-3. Hwy 3, 9.19 km e of St. Thomas, under logs, 4 May 72, JWR, 0-2-0. *Hwy 3, 6.77 km e of Wallacetown, under pine logs, 4 May 72, JWR, 2-0-0. Hwy 3, 3.87 km w of New Glasgow, under lumber in abandoned school yard, 4 May 72, JWR, 3-0-0. ESSEX CO. Hwy 3, 3.87 km w of Leamington, under lumber in dump, 4 May 72, JWR, 10-0-0. FRONTENAC CO. *Hwy 2, 1.13 km e of Westbrook, under logs and paper, 16 May 72, JWR, 5-1-9. *Hwy 15, 4.03 km n of Hwy 401, under paper in wet ditch, 16 May 72, JWR, 0-1-0. *Hwy 41, 1.29 km n of Cloyne, under logs in ditch, 16 May 72, JWR, 0-0-8. GRENVILLE CO. Hwy 2, Cardinal, w.e., under log and paper, 11 May 72, JWR, 2-0-0. GREY CO. *Hwy 6BP, Rockford, w.e., under logs, 5 May 72, JWR, 7-1-0. *Hwy 6, 3.06 km s of Chatsworth, under logs, 5 May 72, JWR, 2-0-0. HALDIMAND CO. *Hwy 3, 9.03 km e of Dunnville, wet ditch, 1 May 72, JWR, 3-0-1. HALIBURTON CO. *Reynolds (1972a). *Hwy 121, 3.22 km e of Tory Hill, dump, 16 May 72, JWR, 6-1-21. HURON CO. *Hwy 83, 6.77 km w of Russeldale, under logs, 5 May 72, JWR, 3-1-0. *Hwy 4, 1.61 km n of Exeter, wet ditch, 5 May 72, JWR, 5-1-0. Hwy 4, 1.29 km n of Hensall, under corn (*Zea maize*) cobs in pasture, 5 May 72, JWR, 1-0-0. *Hwy 4, 3.71 km s of Brucefield, under log and rock, 5 May 72, JWR, 2-0-0. *Hwy 4, 2.74 km n of Clinton, under logs, 5 May 72, 2-1-2. *Hwy 4, 1.29 km s of Wingham, under log in wet area, 5 May 72, JWR, 1-0-0. KENORA DIST. Rushing River Provincial Park, stream trickle in campground, 13-14 Jun 71, ROM Field Party, 1-0-0, ROM-I36. LENNOX AND ADDINGTON CO. *Hwy 500, .81 km n of Denbigh, under log, 16 May 72, JWR, 1-0-0. Hwy 7, Kaladar, e.e., under lumber in school yard, 16 May 72, JWR, 2-0-2. MANITOULIN DIST. *Hwy 68, 5.97 km n of Birch Island, under rock, 13 May 72, JWR & JEM, 0-0-1. *Hwy 68, 2.42 km s of Birch Island, under logs, 13 May 72, JWR & JEM, 3-1-22. MIDDLESEX CO. *Hwy 73, .97 km n of Mossley, under logs, 4 May 72, JWR, 6-1-6. MUSKOKA DIST. *Hwy 11, 12.9 km s of Bracebridge, under rock, 9 May 72, JWR, 1-0-0. *Hwy 11, 9.19 km s of Hwy 516, under logs and rocks in wet ditch, 9 May 72, JWR, 3-1-0. NIAGARA CO. *Queen Elizabeth Hwy, 1.13 km w of Ontario St, Grimsby, under logs and lumber, 1 May 72, JWR, 4-1-0. *Niagara Co. Rd 12, 3.06 km s of Grimsby, under paper in wet ditch, 1 May 72, JWR, 5-1-6. *Hwy 20, 1.61 km e of Smithville, old house site under lumber, 1 May 72, JWR, 4-1-1. NIPISSING DIST. *Hwy 17, 5.48 km e of Warren, under log and dung in pasture, 13 May 72, JWR & JEM, 5-5-10. Hwy 17, 15.81 km e of Sturgeon Falls, dump, 13 May 72, JWR & JEM, 6-4-5. NORFOLK CO. *Hwy 6, 4.84 km s of Jarvis, under logs, 1 May 72, JWR, 1-1-1. OXFORD CO. *Hwy 97, Washington, w.e., under log, 3 May 72, JWR, 1-0-0. PARRY SOUND DIST. *Hwy 520, 10.32 km e of Magnetawan, under lumber, 9 May 72, JWR, 5-2-3. *Hwy 520, Magnetawan, e.e., under rocks, 9 May 72, JWR, 2-0-1. *Hwy 124, 2.26 km w of Hwy 520, under logs, 9 May 72, JWR, 4-3-2. *Hwy 124, 3.06 km e of McKellar, under logs, 9 May 72, JWR, 0-0-3. *Hwy 124, 3.55 km n of Hwy 69, under logs, 9 May 72, JWR, 1-0-3. PETERBOROUGH CO. *Hwy 28, Lakefield College, under logs near waterfront, 27 Apr 72, JWR & CWR, 2-0-0. *Hwy 504, 1.61 km e of Apsley, under log, 27 Apr 72, JWR, 1-0-0. RENFREW CO. *4.19 km s of Palmer Rapids, under paper in ditch, 16 May 72, JWR, 2-2-3. RUSSELL CO. *Hwy 17, 7.74 km e of Rockland, under paper in wet ditch, 11 May 72, JWR, 9-1-0. *Hwy 17, 4.35 km w of Rockland, under logs, 11 May 72, JWR, 9-2-4. *2.42 km s of Limonges, under lumber at old house site, 11 May 72, JWR, 0-1-0. *2.9 km e of Embrun, under logs, 11 May 72, JWR, 3-0-3. SIMCOE CO. Hwy 27, 3.22 km s of Newton Robinson, wet ditch, 2 May 72, JWR, 2-0-0. *Barrie, park opposite 14 Greenfield, under rock and grass clippings, 7 May 72, JWR, 6-1-0. STORMONT CO. *Hwy 43, 5.48 km w of Finch, in and under straw in wet ditch, 11 May 72, JWR, 3-1-0. SUDBURY DIST. *Hwy 69, 3.39 km n of Estaire, under rocks and grass near house, 13 May 72, JWR & JEM, 3-0-6. *Hwy 17, 2.74 km e of Whitefish, under paper in ditch, 13 May 72, JWR & JEM, 3-2-8. VICTORIA CO. *Hwy 7, 1.94 km w of Hwy 46, under logs and dung in pasture, 9 May 72, JWR, 5-0-0. Hwy 48, Bolsover, e.e., under paper in wet ditch, 9 May 72, JWR, 0-1-0. WATERLOO CO. Hwy 86, West Montrose, under log, 3 May 72, JWR, 1-0-0. Waterloo, Beechwood South, on sidewalk after rain (evening), 15 Jun 75, DPS, 2-0-1, UW-0004. WELLINGTON CO. *Hwy 24, 5.16 km e of Eramosa, under log in cedar (*Thuja occidentalis*) woodlot, 29 Apr 72, JWR, 1-0-0. WENTWORTH CO. Reynolds (1972a). *Hwy 6, 5.32 km n of Hwy 5, under logs, 29 Apr 72, JWR, 2-0-1. YORK CO. Edenbrook Park, Islington, quantitative study 1, formalin, 18 May 72, JWR, 34-0-0. Edenbrook Park, Islington, quantitative study 2, formalin, 18 May 72, JWR, 47-0-1.

Lumbricus terrestris Linnaeus, 1758

Nightcrawler, Dew-worm Ver nocture rampant

(Fig. 34)

1758 *Lumbricus terrestris* (part.) Linnaeus, Syst. Nat. (ed. 10), p. 647.
1774 *Lumbricus terrestris* (part.)–Müller, Verm. Terr. Fluv. 1(2): 24.
1780 *Lumbricus terrestris* (part.) + *L. norvegicus* (part.)–Fabricius, Fauna Grønlandica, p. 277.
1825 *Lumbricus terrester* (part.)–Blumenbach, Hand. Naturg. (ed. 11), p. 365.
1826 *Enterion herculeum* Savigny, Mém. Acad. Sci. Inst. Fr. 5: 180.
1837 *Lumbricus herculeus*– Dugès, Ann. Sci. Nat., ser. 2, 8: 17, 21.
1842 *Lumbricus agricola* Hoffmeister, Verm. Lumbric., p. 24.
1867 *Lumbricus infelix* Kinberg, Öfv. Vet.-Akad. Förh. Stockholm 23: 98.
1872 ? *Lumbricus americanus* E. Perrier, Nouv. Arch. Mus. Paris 8: 44.
1884 *Lumbricus herculeus*–Rosa, Lumbric. Piemonte, p. 22.
1896 *Lumbricus studeri* Ribaucourt, Rev. Suisse Zool. 4: 5.
1900 *Lumbricus terrestris*–Michaelsen, Das Tierreich, Oligochaeta 10: 511.
1937 *Lumbricus herculeus*–Tétry, Bull. Mus. Hist. Nat. 9: 151.
1953 *Lumbricus terrestris*–Graff, Zool. Anz. 161(11/12): 324.
1958 *Lumbricus terrestris*–Gates, Breviora, Mus. Comp. Zool. 91: 8.
1969 *Lumbricus herculeus*–Bouché, Pedobiologia 9: 89.
1970 *Lumbricus herculeus*–Bouché, Rev. Écol. Biol. Sol 7(4): 541.
1972 *Lumbricus herculeus*–Bouché and Beugnot, Rev. Écol. Biol. Sol 9(4): 697.
1972 *Lumbricus herculeus*–Bouché, Inst. Natn. Rech. Agron., p. 352.
1973 *Lumbricus terrestris*–Sims, Bull. Zool. Nomencl. 30(1): 27.

NOTE: These are the major synonyms and references for *Lumbricus terrestris* Linnaeus. Unfortunately, *Lumbricus terrestris* has become almost synonymous with "earthworm" in many parts of the world. This has led to many published studies and reports about the species which in fact were undertaken on different, and frequently distantly related, species (Örley, 1881; Vaillant, 1889; Stephenson, 1930: xi; Causey, 1952; Stebbings, 1962: 905; Cameron and Fogal, 1963; Gates, 1972a: 114-115; and Gates, 1972c: 120-123).

Diagnosis

Length 90–300 mm, diameter 6–10 mm, segment number 120–160, prostomium tanylobic, first dorsal pore 7/8. Citellum xxxi, xxxii–xxxvii. Tubercula pubertatis xxxiii–xxxvi. Setae enlarged and widely paired in the caudal and cephalic regions (i.e., *AB* and *CD* are greater) but closely paired and smaller in the central region, $AA > BC$, $AB > CD$, and $DD = \frac{1}{2}C$ anteriorly, $DD < \frac{1}{2}C$ posteriorly. The ventral setae of x, xxvi, and sometimes xxv are on broad genital tumescences modified into genital setae. Genital tumescences, occasionally in viii–xiv and xxiv–xxxix. Male pores prominent with large elevated glandular papillae extending over xiv–xvi. Seminal vesicles, three pairs in 9, 11, and 12+13. Spermathecae, two pairs with short ducts opening at 9/10 and 10/11. Body cylindrical and strongly compressed dorsoventrally posteriorly. Heavily pigmented, brownish-red, or violet colour on dorsum and yellowish-orange on venter.

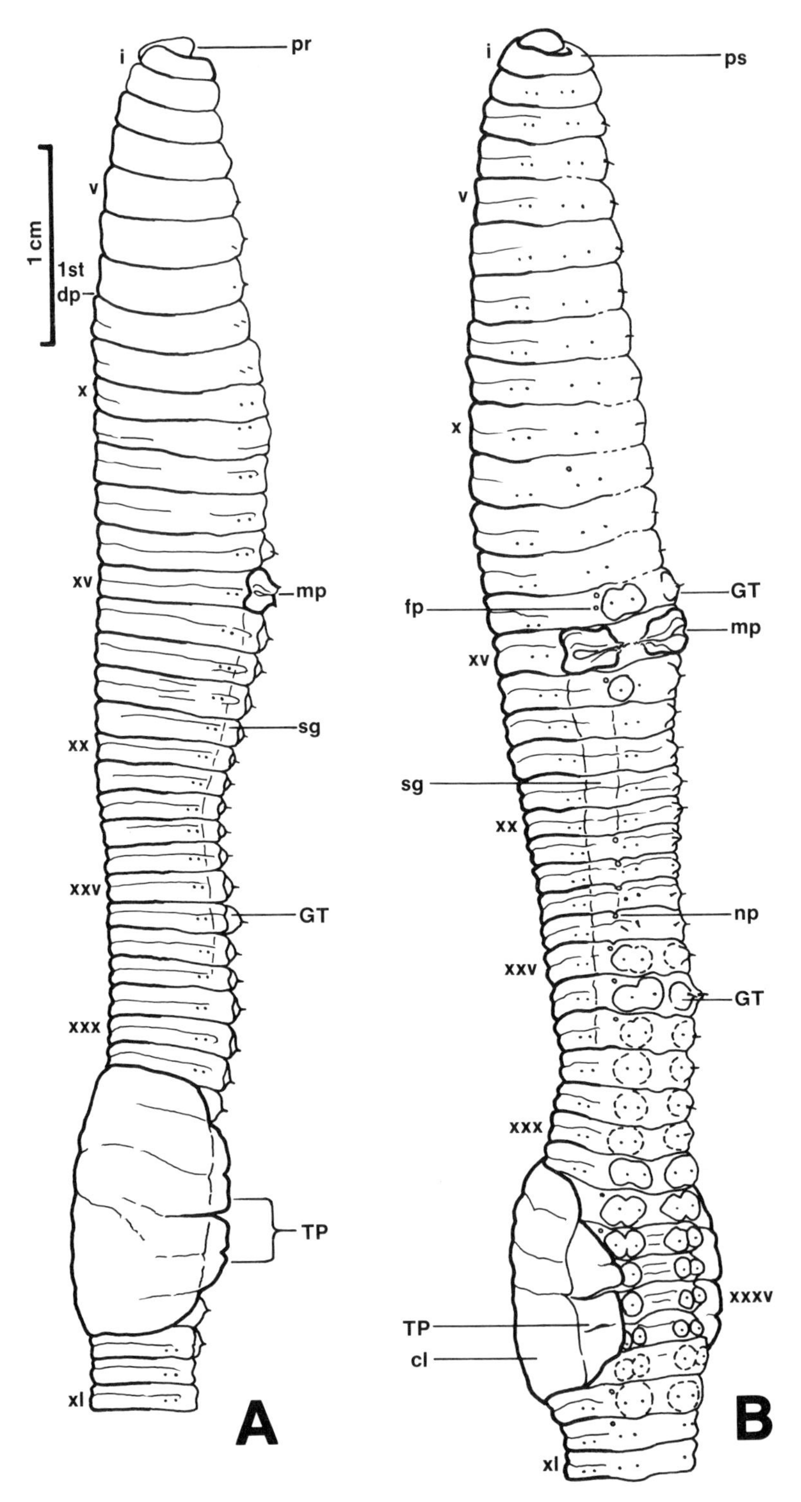

Fig. 34 External longitudinal views of *Lumbricus terrestris* showing taxonomic characters. A. Dorsolateral view. B. Ventrolateral view. (ONT: Manitoulin Dist., cat. no. 7956)

Biology

Lumbricus terrestris has been found in soils of pH 4.0–8.08 and can adapt to a wide variety of habitats. According to Gerard (1964) it is almost purely terrestrial and is found in gardens, arable and pasture lands, forests, and river banks. Gates (1972c) records additional habitats such as streams, mud flats, woody peat, under logs in a stream bed, under cow pats, and in compost. It has been found in greenhouses and botanical gardens in Europe and North America and in European caves (Gates, 1972c). The species does not normally occur in forests in North America (Reynolds, 1976b). After intensive collecting by the author in eastern North America, it now appears that statements such as "*Lumbricus terrestris* is becoming increasingly important in the United States, following its importation and gradual replacement of endemic populations" (Edwards et al., 1969) are not true. Many erroneous statements such as this have been made about *L. terrestris* without data to support their claims. Data for naturally occurring populations of *L. terrestris* in North American areas south of the limits of Quaternary glaciations are lacking (Gates, 1970). A few limited collections of this species were obtained in Tennessee, Maryland, and Delaware, three states south of the southern limits of Pleistocene glaciation, by Reynolds et al. (1974), Reynolds (1974b), and Reynolds (1973a), respectively. In Ontario, *L. terrestris* was collected from a variety of habitats, but primarily from under logs. Millions are collected annually from the surfaces of golf courses and lawns for the bait industry.

Under favourable conditions activity, including copulation, is year round but a summer and winter rest period may be climatically imposed in certain areas. The species is obligatorily amphimictic (Reynolds, 1974c) and copulation is nocturnal and takes place on the surface of the soil. Feeding may be both selective and indiscriminate and certainly during burrowing much soil is swallowed. Casting is usually beneath the soil surface. Unusual activities of this species are the lining of burrows with pebbles, or faecal earth, and the drawing into the burrows of leaves. The entrances to burrows may be blocked with seeds, sticks, straws, and feathers.

Lumbricus terrestris is an important species in litter decomposition. It is also collected annually at night by the millions as the major species of worm for fish bait in the northern portion of North America. The relatively long life cycle makes it unprofitable to rear this species commercially for fish bait (cf. *Aporrectodea tuberculata* and *Eisenia foetida*). It is also gathered annually in large numbers for biological supply institutions which distribute this species for use in laboratory studies at high schools and universities, for *L. terrestris* is nearly always the textbook example of the oligochaetes. For example, 20,000 specimens per year are shipped to South Africa for biological studies because it cannot be found locally (Reinecke, in litt.).

Range

A native of Palaearctis, *L. terrestris* is now known from Europe, Iceland, North America, South America, Siberia, South Africa, and Australasia (Gates, 1972c). Reinecke (in litt.), however, cannot confirm its natural occurrence in South Africa.

North American Distribution

British Columbia (Berkeley, 1968), New Brunswick (Reynolds, 1976d), Newfoundland (Smith, 1917), Nova Scotia (Reynolds, 1975a, 1976a), Ontario (Reynolds, 1972a), Prince Edward Island (Reynolds, 1975c), Québec (Reynolds, 1975d, e, 1976c), Arkansas (Causey, 1952), California (Smith, 1917), Colorado (Smith, 1917), Connecticut (Reynolds, 1973c), Delaware (Reynolds, 1973a), District of Columbia (Smith, 1917), Hawaii (Reynolds et al., 1974), Idaho (Gates, 1967), Illinois (Smith, 1917), Indiana (Joyner, 1960), Iowa (Gates, 1967), Kansas (Gates, 1967), Maine (Smith, 1917), Maryland (Reynolds, 1974b), Massachusetts (Reynolds, 1977), Michigan (Smith, 1917), Minnesota (Gates, 1972c), Missouri (Olson, 1936), New Hampshire (Reynolds, 1976c), New Jersey (Davies, 1954), New York (Olson, 1940), North Carolina (Gates, 1972c), Ohio (Olson, 1928), Oregon (MacNab and McKey-Fender, 1947), Pennsylvania (Bhatti, 1965), Rhode Island (Reynolds, 1973b), Tennessee (Reynolds, 1974a), Utah (Gates, 1967), Vermont (Gates, 1949), Virginia (Reynolds, 1974b), Washington (Altman, 1936), West Virginia (Reynolds, 1974b), Wisconsin (Gates, 1972c), Greenland (Eisen, 1872). New records: Georgia.

Ontario Distribution (Fig. 35)

Lumbricus terrestris was first reported from Ontario by Stafford (1902), and later by Judd (1964) and Reynolds (1972a). It is widely distributed over the province, perhaps, in part, as a result of the fish bait industry.

ALGOMA DIST. *Hwy 17, .48 km e of Spanish, under log and paper in ditch, 13 May 72, JWR & JEM, 0-2-0. BRANT CO. *Hwy 53, 4.35 km e of Cathcart, under log, 1 May 72, JWR, 1-0-0. *Hwy 54, 5.16 km e of Middleport, ditch, 3 May 72, JWR, 4-0-2. *Hwy 2, 6.45 km e of Paris, under lumber, 3 May 72, JWR, 0-0-1. New Durham, May 30, RC, 1-1-2, ROM. BRUCE CO. *Hwy 4, 6.77 km s of Teeswater, under logs, 5 May 72, JWR, 0-0-4. *Hwy 4, .48 km s of Hwy 9, Under logs, 5 May 72, JWR, 0-2-0. CARLETON CO. Reynolds (1972a). COCHRANE DIST. Reynolds (1972a). DUFFERIN CO. *Hwy 10-24, 8.06 km n of Orangeville, under logs, 2 May 72, JWR, 1-1-1. *Hwy 10-24, 7.74 km n of Camilla, under logs, 2 May 72, JWR, 9-1-0. DURHAM CO. *Hwy 401, 1.61 km e of Newtonville, under logs, 15 May 72, JWR, 4-0-1. ELGIN CO. *Hwy 3, 2.1 km e of Wallacetown, under log, 4 May 72, JWR, 0-0-1. FRONTENAC CO. *Hwy 15, 4.68 km s of Seeley's Bay, under logs, 16 May 72, JWR, 1-1-0. GLENGARRY CO. *Hwy 401, 7.45 km w of Summerstown Rd, under log, 11 May 72, JWR, 0-0-1. GRENVILLE CO. *Hwy 2, 5.16 km w of Prescott, under rock, 10 May 72, JWR, 0-0-1. *Hwy 2, Johnstown, w.e., under logs, 11 May 72, JWR, 0-1-1. GREY CO. *Hwy 21, 7.26 km w of Springmount, under rocks, 5 May 72, JWR, 9-0-1. *Hwy 10, 8.06 km s of Flesherton, under logs, 5 May 72, JWR, 0-0-2. HALDIMAND CO. *Hwy 3, 9.03 km e of Dunnville, wet ditch, 1 May 72, JWR, 0-0-1. *Hwy 3, 6.29 km w of Dunnville, 1 May 72, JWR, 0-0-1. *Hwy 54, 2.26 km n of Caledonia, digging, 3 May 72, JWR, 1-0-1. HALIBURTON CO. *Reynolds (1972a). *Hwy 121, 3.22 km e of Tory Hill, dump, 16 May 72, JWR, 0-1-3. HALTON CO. *Halton Co. Rd 3, Esquesing Community, under rocks and logs, next to the Town Hall, 4 May 73, JWR, 4-0-0. *Hwy 7, 4.84 km e of Georgetown, under burnt log, 4 May 73, JWR, 1-0-0. HASTINGS CO. Reynolds (1972a). *Hwy Jct 2 and 49, wet ditch, 15 May 72, JWR, 1-2-1. HURON CO. *Hwy 4, 2.74 km n of Clinton, under logs, 5 May 72, JWR, 0-1-1. KENT CO. *River Rd, 6.45 km w of Chatham, under logs, 23 Apr 72, JWR & TW, 6-0-3. *Hwy 3, 9.03 km e of Port Alma, under log, 4 May 72, JWR, 0-0-1. *Hwy 3, 3.55 km e of Wheatley, digging, 4 May 72, JWR, 0-0-1. LAMBTON CO. Hwy 21, .65 km s of Petrolia, ditch, 4 May 72, JWR, 1-1-0. LANARK CO. *Hwy 43, 3.71 km e of Smiths Falls, under telephone pole in ditch, 11 May 72, JWR, 0-1-3. LEEDS CO. *Hwy 15, 2.26 km n of Seeley's Bay, under log and concrete block in ditch, 16 May 72, JWR, 0-0-2. LENNOX AND ADDINGTON CO. *Hwy 2, 9.68 km w of Napanee, under paper in ditch, 15 May 72, JWR, 0-0-1. *Hwy 2, 6.61 km w of Hwy 133, under logs, 16 May 72, JWR, 0-1-1. MANITOULIN DIST. *Hwy 68, Birch Island, s.e., under logs, 13 May 72, JWR & JEM, 2-1-0. *Hwy 68, 3.22 km n of Little Current, under log, 13 May 72, JWR & JEM, 0-0-1. MIDDLESEX CO. Judd (1964). *Hwy 7, 2.26 km s of Parkhill, under fence posts, 4 May 72, JWR, 2-0-1. MUSKOKA DIST. *Hwy 11, 11.45 km n of Severn Bridge, under pine log, 9 May 72, JWR, 0-1-0. NIPISSING DIST. *Hwy 17, 8.39 km e of Warren, house site under lumber, 13 May 72, JWR & JEM, 1-1-1. NORFOLK CO. *Hwy 6, 4.84 km s of Jarvis, under log, 1 May 72, JWR, 0-0-1. *Hwy 3, 11.29 km e of Delhi, under log, 1 May 72, JWR, 0-0-1. NORTHUMBERLAND CO. *Hwy 401, 5.65 km e of Grafton, under logs, 15 May 72, JWR, 0-0-2. ONTARIO CO. *Hwy 7, 1.61 km e of Green River, under log, 26 Apr 72, JWR, 1-0-0. *Hwy 47, 10.48 km w of Uxbridge, digging, 9 May 72, JWR, 0-1-0. OXFORD CO. *Hwy 59, 1.77

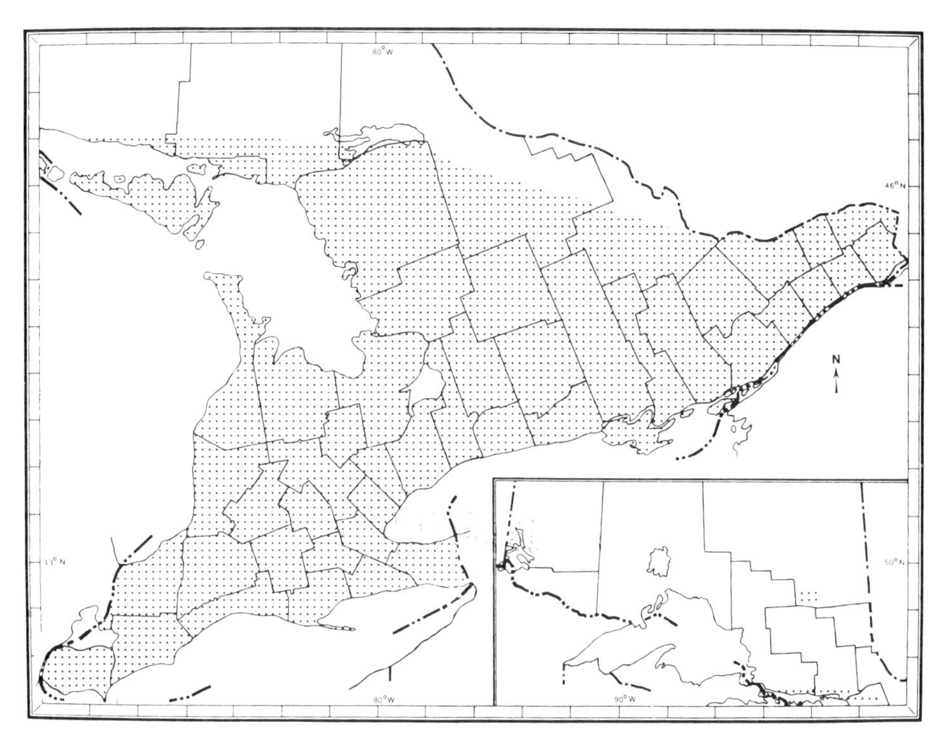

Fig. 35 The known Ontario distribution of *Lumbricus terrestris*.

km e of Burgessville, under log, 1 May 72, JWR, 0-0-1. *Hwy 59, .32 km n of Hickson, under logs, 3 May 72, JWR, 2-1-1. PARRY SOUND DIST. *Hwy 11, 4.52 km s of Burks Falls, under log, 9 May 72, JWR, 0-0-1. PEEL CO. *Hwy 5, 4.35 km w of Hwy 10, under mattress, 29 May 72, JWR, 0-0-2. *Hwy 24, 8.87 km n of Erin, under log in pasture, 29 Apr 72, JWR, 0-0-1. *Hwy 7, Brampton, e.e., under debris, 4 May 73, JWR, 1-1-0. Hwy 401, 1.3 km w of Dixie Rd, wet ditch, 4 May 73, JWR, 0-0-2. PERTH CO. .81 km n of Fullarton, under rocks, 29 Apr 72, DWR & LWR, 4-2-0. *Mitchell, next to Collegiate, under logs, 4 May 72, JWR & DWR, 3-0-2. Hwy Jct 23 and 83, Russeldale, under paper in wet ditch, 5 May 72, JWR, 0-0-1. PETERBOROUGH CO. *Hwy 28, 5.48 km n of Burleigh Falls, under log, 27 Apr 72, JWR, 0-1-0. PRINCE EDWARD CO. *Hwy 33, 1.61 km s of Carrying Place, under log in wet ditch, 15 May 72, JWR, 0-0-1. *Hwy 33, 2.58 km e of Hillier, digging, 15 May 72, JWR, 0-0-1. Hwy 33, Bloomfield, e.e., under paper, 15 May 72, JWR, 0-0-1. RUSSELL CO. *2.9 km e of Embrun, under logs, 11 May 72, JWR, 0-1-1. RENFREW CO. Reynolds (1972a). SIMCOE CO. *Hwy 27, 5.97 km s of Cookstown, under logs, 2 May 72, JWR, 3-1-0. *Hwy 89, 1.94 km e of Rosemont, under log in wet ditch, 2 May 72, JWR, 0-0-1. *Hwy 89, 5.48 km e of Alliston, under log, 2 May 72, JWR, 1-0-0. Barrie, 14 Greenfield, digging in garden and under logs, 7 May 72, JWR & MKG, 6-0-2. *Hwy 400, .81 km s of Hwy 103, under logs, 9 May 72, JWR, 1-1-1. SUDBURY DIST. *Hwy Jct 64 and 69, under rocks, 13 May 72, JWR & JEM, 1-2-1. VICTORIA CO. *Hwy 46, 3.71 km n of Hwy 7, under logs, 9 May 72, JWR, 1-0-2. Hwy 48, Bolsover, e.e., under paper in wet ditch, 9 May 72, JWR, 1-0-0. WATERLOO CO. *Hwy 24A, 11.45 km n of Paris, under log, 3 May 72, JWR, 1-0-1. *Hwy 97, 3.71 km w of Hwy 401, under logs, 3 May 72, JWR, 1-0-1. WELLINGTON CO. *Hwy 24, 4.03 km n of Erin, under log, 29 Apr 72, JWR, 0-1-0. Hwy 6, University of Guelph, on wet driveway behind the Soil Science Building, 2 May 72, JWR, 0-1-3. *Hwy 6, 1.13 km s of Ennotville, under paper in wet ditch, 2 May 72, 0-0-1. *Hwy 6, 10.64 km s of Arthur, under logs, 2 May 72, JWR, 1-0-2. WENTWORTH CO. Reynolds (1972a). *Hwy 5, Waterdown, e.e., edge of corn (*Zea maize*) field, 29 Apr 72, JWR, 1-0-0. YORK CO. *Hwy 48, .32 km n of Steeles Avenue, under log, 26 Apr 72, JWR, 0-1-0. *Edenbrook Park, Islington, under logs and rocks near stream bank, 30 Apr 72, JWR & DWR, 34-1-1. Islington, 12 Country Club Drive, crawling on grass, 13 May 72, RGR & WMR, 0-0-2. Edenbrook Park, Islington, quantitative study 1, formalin, 18 May 72, JWR, 1-0-2. Edenbrook Park, Islington, quantitative study 2, formalin, 18 May 72, JWR, 7-0-3. Scarborough Bluffs, Brimley Rd, garden, 28 Oct 41, JO, 1-0-8, ROM.

Genus *Octolasion* Örley, 1885

1885 *Octolasion* (part.) Örley, Ertek. Term. Magyar Akad. 15(18): 13.
1889 *Lumbricus (Octolasion)* (part.) + *L. (Allobophora)* (part.) + *Dendrobaena* (part.) + *L. (L.)* (part.) + *Titanus* ? (part.)–L. Vaillant, Hist. Nat. Annel. 3(1): 113, 130, 116, 121, 93.
1896 *Octolasion* + *Allolobophora (Octolasion)*–Ribaucourt, Rev. Suisse Zool. 4: 95.
1900 *Octolasium* (part.)–Michaelsen, Das Tierreich, Oligochaeta 10: 504.
1930 *Octolasium* (part.)–Stephenson, Oligochaeta (Oxford), p. 914.
1972 *Octolasium*–Bouché, Inst. Natn. Rech. Agron., p. 253.
1975 *Octolasion*–Gates, Megadrilogica 2(1): 4.

Type Species

Octolasion lacteum Örley, 1885 (= *Enterion tyrtaeum* Savigny, 1826).

Diagnosis

Calciferous sacs, in x, large, lateral, communicating vertically and widely with gut lumen though reaching beyond oesophagus both dorsally and ventrally. Calciferous lamellae continued onto posterior walls of sacs. Intestinal origin, in xv. Gizzard, mostly in xvii. Extraoesophageals, passing up to dorsal trunk posteriorly in xii. Hearts, vi–xi. Nephridial bladders, ocarina-shaped, Nephropores, obvious, behind xv in one regular rank on each side, just above *B*. Setae, behind the clitellum not closely paired. Prostomium epilobic. Longitudinal musculature, pinnate. (after Gates, 1972c: 123, 1975a: 4).

Discussion

There has been considerable confusion concerning the spelling of this genus name since the early 1900s. Michaelsen (1900b) changed many of the Greek generic endings to Latin endings, i.e., *Octolasion* to *Octolasium* and *Bimastos* to *Bimastus*. According to the International Code of Zoological Nomenclature (Article 32), the original spelling is the correct spelling. Therefore, *Octolasion* Örley, 1885 and *Bimastos* Moore, 1893 are the correct spellings and most current oligochaetologists are now employing them. The genus *Octolasion* Örley, 1885 contains only two species, *O. cyaneum* and *O. tyrtaeum*. Species in Europe and elsewhere often placed in *Octolasion* are now considered to belong in the genus *Octodrilus* Omodeo, 1956 with *Lumbricus complanatus* Dugès, 1828 as the type (cf. Bouché, 1972, Gates, 1975a).

Octolasion cyaneum (Savigny, 1826)

Woodland blue worm   Ver bleu des bois

(Fig. 36)

1774 *Lumbricus terrestris* (part.)–Müller, Verm. Terr. Fluv. 1(2): 24.

1826 *Enterion cyaneum* Savigny, Mém. Acad. Sci. Inst. Fr. 5: 181.

1837 *Lumbricus cyaneus*–Dugès, Ann. Sci. Nat., ser. 2, 8: 17, 21.

1845 *Lumbricus stagnalis* (part.) Hoffmeister, Regenwürmer, p. 35.

1867 *Lumbricus alyattes* Kinberg, Öfv. Vet.-Akad. Förh. Stockholm 23: 99.

1889 *Lumbricus alyattes* + *Lumbricus (Dendrobaena) stagnalis* + *Lumbricus cyaneus*–L. Vaillant, Hist. Nat. Annel. 3(1): 96, 118, 124.

1890 *Allolobophora studiosa* Michaelsen, Arch. Ver. Mechlenb. 44: 50.

1893 *Allolobophora (Octolasion) cyanea* (part.)–Rosa, Mem. Acc. Torino, ser. 2, 43: 424, 455, 456.

1896 *Allolobophora (Octolasion) cyanea studiosa*–Ribaucourt, Rev. Suisse Zool. 4: 95.

1900 *Octolasium cyaneum*–Michaelsen, Das Tierreich, Oligochaeta 10: 506.

1972 *Octolasium cyaneum*–Bouché + *O. c.* var. *armoricum* Bouché, Inst. Natn. Rech. Agron., p. 258, 260.

1972 *Octolasium cyaneum*–Edwards and Lofty, Biol. earthworms, p. 214.

1972 *Octolasion cyaneum*–Gates, Bull. Tall Timbers Res. Stn. 14: 31.

Diagnosis

Length 65–180 mm, diameter 7–8 mm, segment number 140–158, prostomium epilobic, first dorsal pore 11/12 or 12/13. Clitellum xxix–xxxiv. Tubercula pubertatis xxx–xxxiii. Setae closely paired anteriorly, $CD<AB<BC<AA<DD$, and widely paired posteriorly, $AB>BC>CD$. Setae of x, xviii, xix, xx, xxi frequently on white genital tumescences. Male pores on xv with well-defined narrow papillae. Seminal vesicles, four pairs in 9–12, with the pairs in 11 and 12 larger than the pairs in 9 and 10. Spermathecae, two pairs opening between *c* and *d* in 9/10 and 10/11. Body cylindircal but octagonal posteriorly. Colour, blue-grey or whitish.

Biology

This species is known from soils of pH 5.2–8.0 and may well be ubiquitous with respect to this factor. Gates (1972c) records it from under stones in water, in moss, stream banks, and other limnic habitats. It is known also from ploughed fields, wet sand, forest soils, and from caves in Europe. Černosvitov and Evans (1947) reported it mostly from under stones and occasionally under moss. Gates (1973a) found the species under logs and under rocks near stream beds. Under logs and rocks was the most common site for *O. cyaneum* in Ontario.

Activity may be year round but in central Maine summer drought and winter freezing impose two periods of inactivity (Gates, 1972c), and this is probably the case in Ontario. In experimental studies casting was below ground but occasional surface casting has been reported (Gates, 1972c). *O. cyaneum* is obligato-

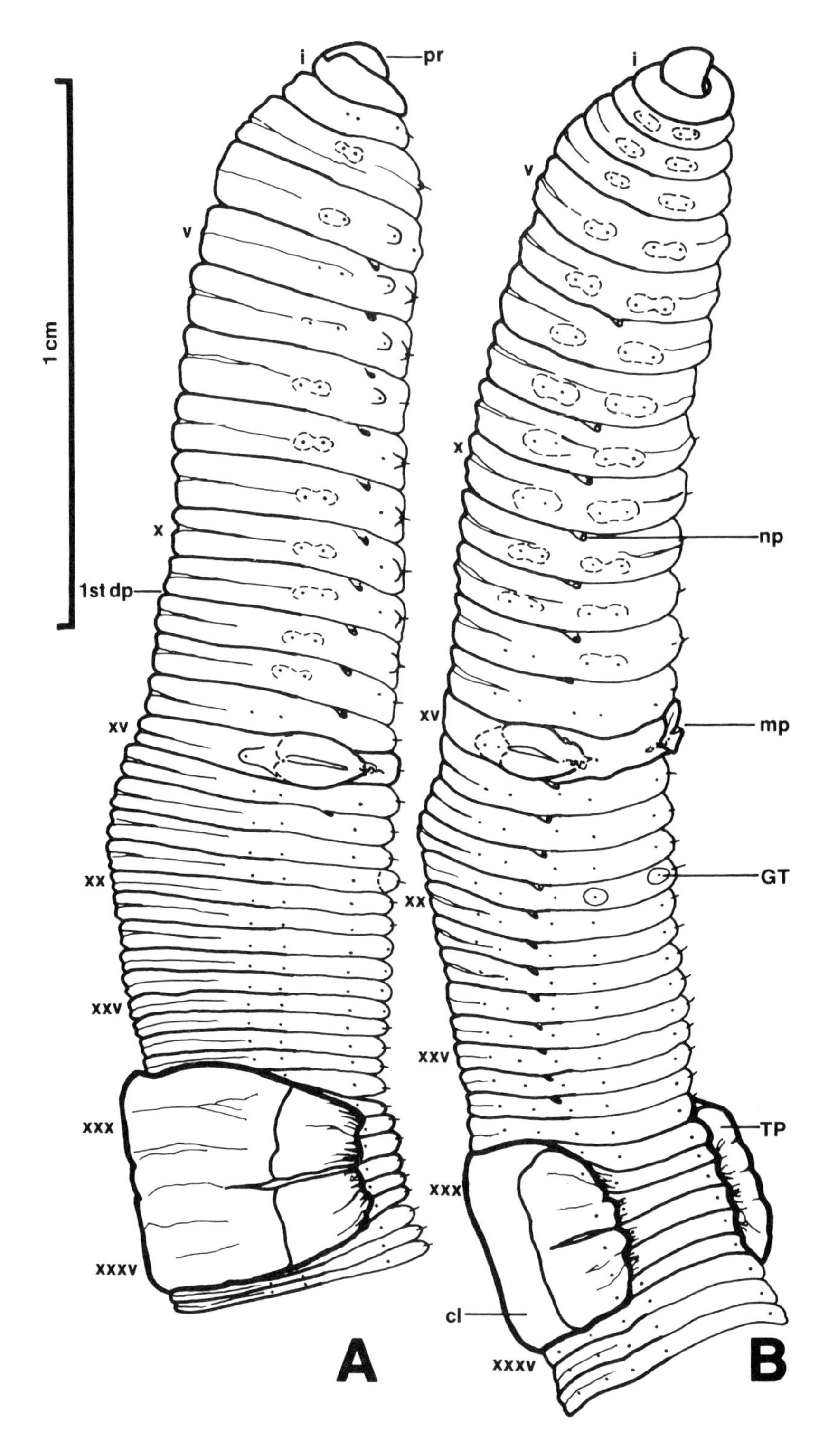

Fig. 36 External longitudinal views of *Octolasion cyaneum* showing taxonomic characters. A. Lateral view. B. Ventrolateral view. (ONT: Durham Co., cat. no. 3384)

rily parthenogenetic (Reynolds, 1974c); copulation has not been recorded and may never have been observed (Gates, 1972c). This species is relatively rare in North America and is of little or no economic importance.

Range

A native of Palaearctis, *O. cyaneum* is now known from Europe, Iceland, North America, South America, India, Azores, and Australasia (Gates, 1972c).

North American Distribution

British Columbia (Wickett, 1967), Ontario (Reynolds, 1972a), Nova Scotia (Reynolds, 1976a), California (Gates, 1966), Colorado (Gates, 1967), Georgia (Gates, 1973a), Indiana (Reynolds et al., 1974), Iowa (Evans, 1948a), Maine (Gates, 1966), Massachusetts (Reynolds, 1977), New York (Schwert, 1976), North Carolina (Gates, 1973a), Pennsylvania (Reynolds, 1974b), Tennessee (Reynolds, 1974a), Virginia (Gates, 1973a), Washington (Gates, 1973a). New records: Québec, Mississippi.

Ontario Distribution (Fig. 37)

Octolasion cyaneum was first reported from Ontario by Reynolds (1972a) and the report presented here is only the second account of this species in Ontario. *O. cyaneum* has been obtained only from Haliburton and Perth counties.

HALIBURTON CO. *Reynolds (1972a). PERTH CO. .81 km n of Fullarton, under rocks, 29 Apr 72, DWR & LWR, 1-1-0.

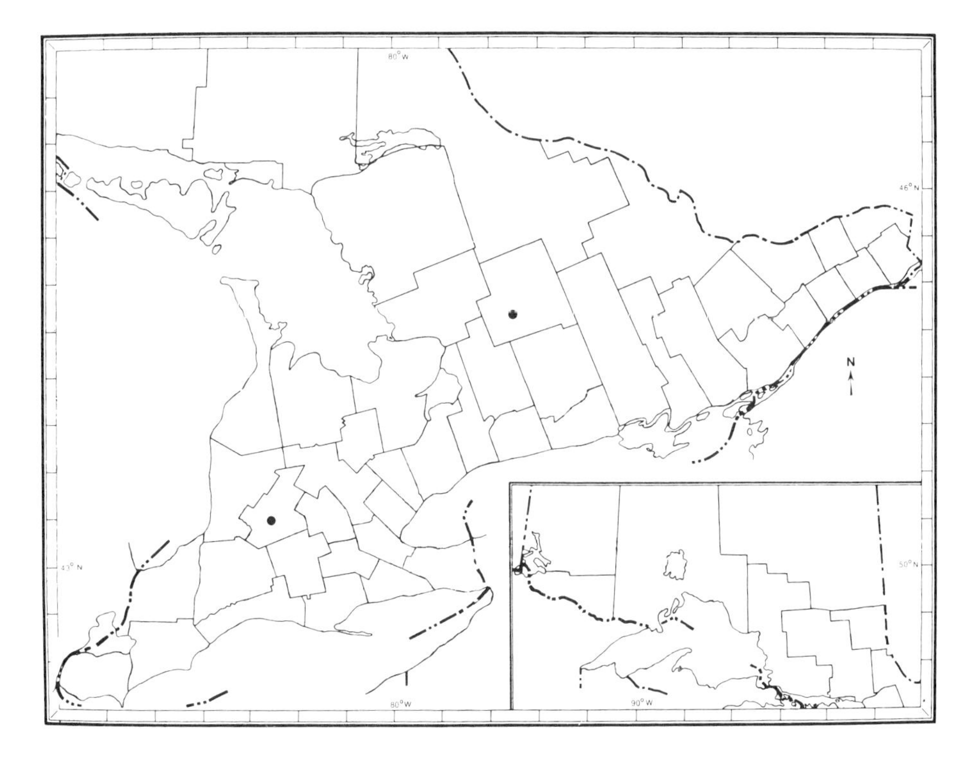

Fig. 37 The known Ontario distribution of *Octolasion cyaneum*.

***Octolasion tyrtaeum* (Savigny, 1826)**

Woodland white worm Ver blanc des bois

(Fig. 38)

1826 *Enterion tyrtaeum* Savigny, Mém. Acad. Sci. Inst. Fr. 5: 180.
1837 *Lumbricus tyrtaeus*–Dugès, Ann. Sci. Nat., ser. 2, 8: 17, 22.
1845 *Lumbricus communis cyaneus* + *L. stagnalis* (part.) Hoffmeister, Regenwürmer, p. 24, 35.
1881 *Lumbricus terrestris* var. *lacteus* + *L. t.* var. *rubidus* Örley, Math. Term. Közlem. Magyar Akad. 16: 584.
1884 *Allolobophora profuga* Rosa, Lumbric. Piemonte, p. 47.
1885 *Octolasion rubidum*–Örley + *O. profugum*–Örley + *O. gracile* Örley + *O. lacteum*–Örley, Ertek. Term. Magyar Akad. 15(18): 16, 17, 18, 21.
1889 *Lumbricus (Allobophora) profugus* + *L. (O.) gracilis* L. Vaillant, Hist. Nat. Annel. 3(1): 113.
1896 *Allolobophora (Octolasion) rubida*–Ribaucourt + *A. (O.) gracilis*–Ribaucourt + *A. sylvestris* Ribaucourt, Rev. Suisse Zool. 4: 63, 65, 67, 95.
1900 *Octolasium lacteum*–Michaelsen, Das Tierreich, Oligochaeta 10: 506.
1900 *Allolobophora (Octolasion) profuga*–Michaelsen, Abh. Nat. Verh. Hamburg 16(1): 16.
1900 *Allolobophora profuga*–Smith, Bull. Illinois St. Lab. Nat. Hist. 5: 441.
1917 *Octolasium lacteum*–Smith, Proc. U.S. Natn. Mus. 52(2174): 178.
1952 *Octolosium ladeum* (laps.)–Goff, Amer. Midl. Nat. 47: 484.
1971 *Octolasium lacteum*–Crossley, Reichle and Edwards, Pedobiologia 11: 71.
1972 *Octolasium lacteum*–Edwards and Lofty, Biol. earthworms, p. 216.
1972 *Octolasium lacteum lacteum* + *O. l. gracile*–Bouché, Inst. Natn. Rech. Agron., p. 253, 257.
1972 *Octolasion tyrtaeum*–Gates, Bull. Tall Timbers Res. Stn. 14: 35.

Diagnosis

Length 25–130 mm, diameter 3–6 mm, segment number 75–150, prostomium epilobic, first dorsal pore 9/10–13/14, usually 11/12. Clitellum xxx–xxxv. Tubercula pubertatis xxxi–xxxiv. Setal pairings as in *Octolasion cyaneum.* Frequently setae *a* and/or *b* on xxii and occasionally on ix–xii, xiv, xvii, xix–xxiii, xxvii, xxxvii, or xxxviii are on genital tumescences and modified into genital setae. Male pores on xv and on large glandular papillae extending over xiv and xvi, occasionally limited to xv. Seminal vesicles, four pairs in 9–12, with pairs in 11 and 12 larger than pairs in 9 and 10. Spermathecae, two pairs opening on level *C* or between *c* and *d* in 9/10 and 10/11. Body cylindrical but slightly octagonal posteriorly. Colour variable, milky white, grey, blue, or pink.

Biology

Reported from soils of pH 5.5–8.08, *O. tyrtaeum* has been found under stones and logs, in peat, leaf mould, compost, forest litter, gardens, cultivated fields and pastures, bogs, stream banks, in springs, and around the roots of submerged vegetation (Gates, 1972c). The species is also known from caves in Europe and

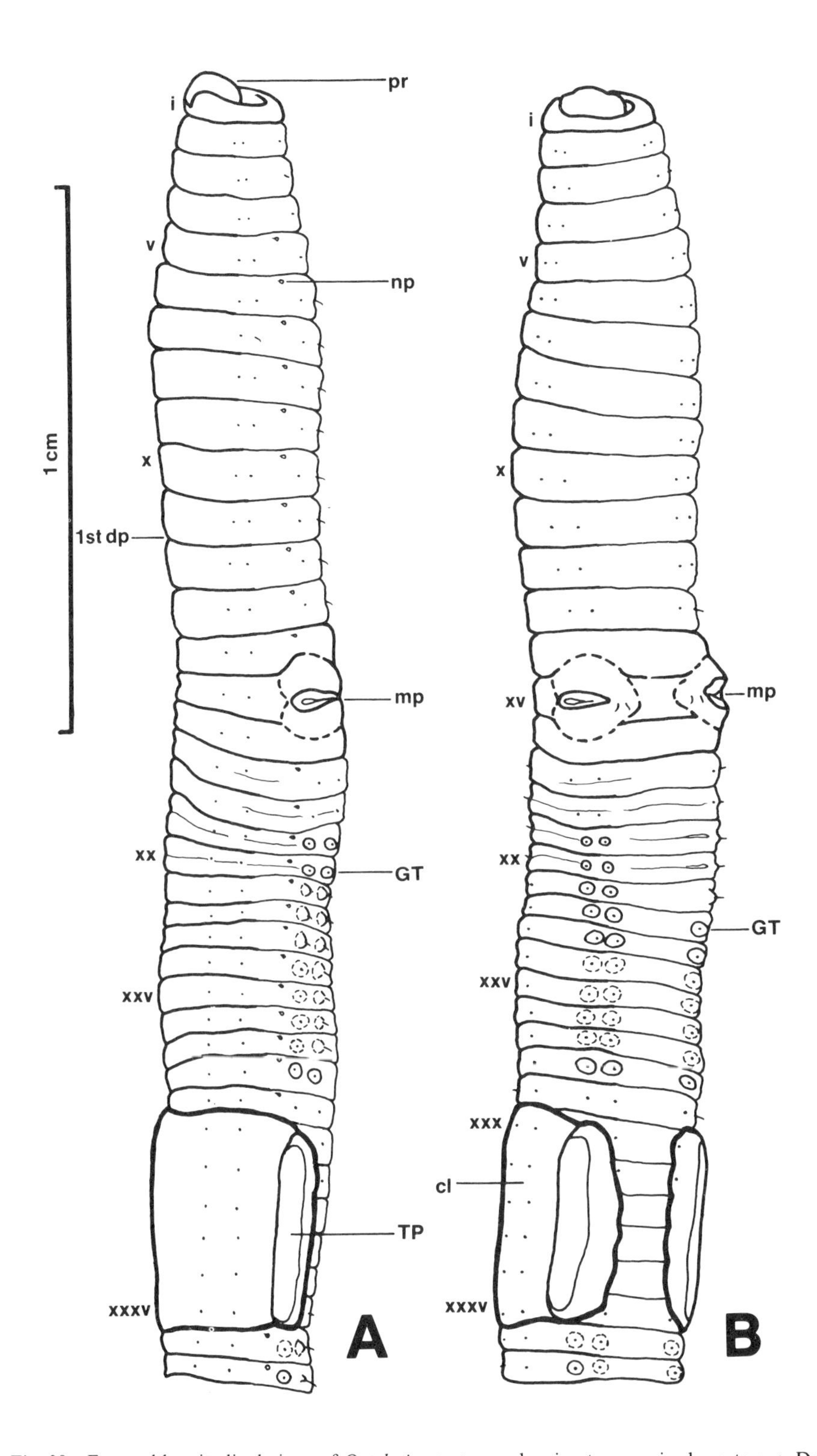

Fig. 38 External longitudinal views of *Octolasion tyrtaeum* showing taxonomic characters. A. Dorsolateral view. B. Ventrolateral view. (ONT: Bruce Co., cat. no. 7557)

North America. Smith (1917) reported this species as commonly found under logs, leaf mould, and debris of various kinds, in compost heaps, and to some extent in the soil. Some workers believed this species preferred rich, moist organic material (Causey, 1952), while others presented data to the contrary (Eaton, 1942). *Octolasion tyrtaeum* was the most abundant species in Tennessee (Reynolds et al., 1974) and was obtained under logs, debris, and rocks, and by digging. In Ontario it was most frequently found under logs.

Activity may be year round although summer drought and winter cold may impose two rest periods. *Octolasion tyrtaeum* is an obligatorily parthenogenetic species (Gates, 1973a; Reynolds, 1974c) and copulation occurs below the surface of the soil. The species is of little economic importance although one dealer in Michigan is reported to have sold it for fish bait (Gates, 1972c).

Range

A native of Palaearctis, *O. tyrtaeum* is known from Europe, North America, South America, Asia, Africa, and Australia (Gates, 1972c).

North American Distribution

British Columbia (Wickett, 1967), Manitoba (Gates, 1972c), Nova Scotia (Reynolds, 1975a, 1976a), Ontario (Reynolds, 1972a), Québec (Reynolds, 1975d, e, 1976c), Alabama (Gates, 1972c), Arkansas (Causey, 1952), California (Smith, 1917), Connecticut (Reynolds, 1973c), Florida (Gates, 1972c), Georgia (Gates, 1972c), Idaho (Gates, 1967), Illinois (Smith, 1917), Indiana (Reynolds, 1972e), Iowa (Gates, 1967), Kentucky (Allee et al., 1930) Maine (Gates, 1961), Maryland (Reynolds, 1974b), Massachusetts (Reynolds, 1977), Michigan (Murchie, 1956), Minnesota (Gates, 1973a), Missouri (Olson, 1936), Nebraska (Gates, 1967), New Jersey (Davies, 1954), New York (Olson, 1940), North Carolina (Gates, 1973a), Ohio (Smith, 1917), Oregon (MacNab and McKey-Fender, 1947), Pennsylvania (Davies, 1954), South Carolina (Gates, 1973a), Tennessee (Reynolds, 1974a), Utah (Gates, 1967), Virginia (Gates, 1973a) West Virginia (Gates, 1959). New records: Alaska, Mississippi, Wisconsin.

Ontario Distribution (Fig. 39)

Octolasion tyrtaeum was first reported from Ontario by Reynolds (1972a) and the present study is only the second report of the species from Ontario. It is more widely distributed than *O. cyaneum* and there are two main centres of distribution, the first around the Kawartha Lakes region in central southern Ontario, and the second in western southern Ontario running along the Niagara Escarpment and to the west.

BRANT CO. *Hwy 54, .64 km w of Onondaga, under log in wet area, 3 May 72, JWR, 0-0-1. BRUCE CO. *Hwy 4, 6.77 km s of Teeswater, under log, 5 May 72, JWR, 0-0-1. *Hwy 4, .48 km s of Hwy 9, under logs, 5 May 72, JWR, 2-2-4. Hwy 21, 2.57 km n of Tiverton, under lumber, 5 May 72, JWR, 9-0-2. DURHAM CO. *Hwy 7A, 2.1 km e of Nestleton Station, under log, 26 Apr 72, JWR, 0-0-1. ELGIN CO. Hwy 3, 9.19 km e of St. Thomas, under logs, 4 May 72, JWR, 2-2-8. GREY CO. *Hwy 6, 6.94 km s of Dornoch, under logs, 5 May 72, JWR, 2-1-2. *Hwy 4, 8.22 km e of Durham, under logs, 5 May 72, JWR, 1-0-1. HALIBURTON CO. *Reynolds (1972a). HALTON CO. Hwy 5, 5.81 km e of Hwy 25, under paper and leaves in ditch, 29 Apr 72, JWR, 1-0-1. *Halton Co. Rd 3, Esquesing Community, under rocks and logs next to the Town Hall, 4 May 73, JWR, 0-4-3-1. Campbellville, rapids in cold stream, 15 May 66, IMS, ST, & RNS, 0-1-0, ROM-I6. HURON CO. *Hwy 83, 6.77 km w of Russeldale, under logs, 5 May 72, JWR, 3-0-1. NORTHUMBER-

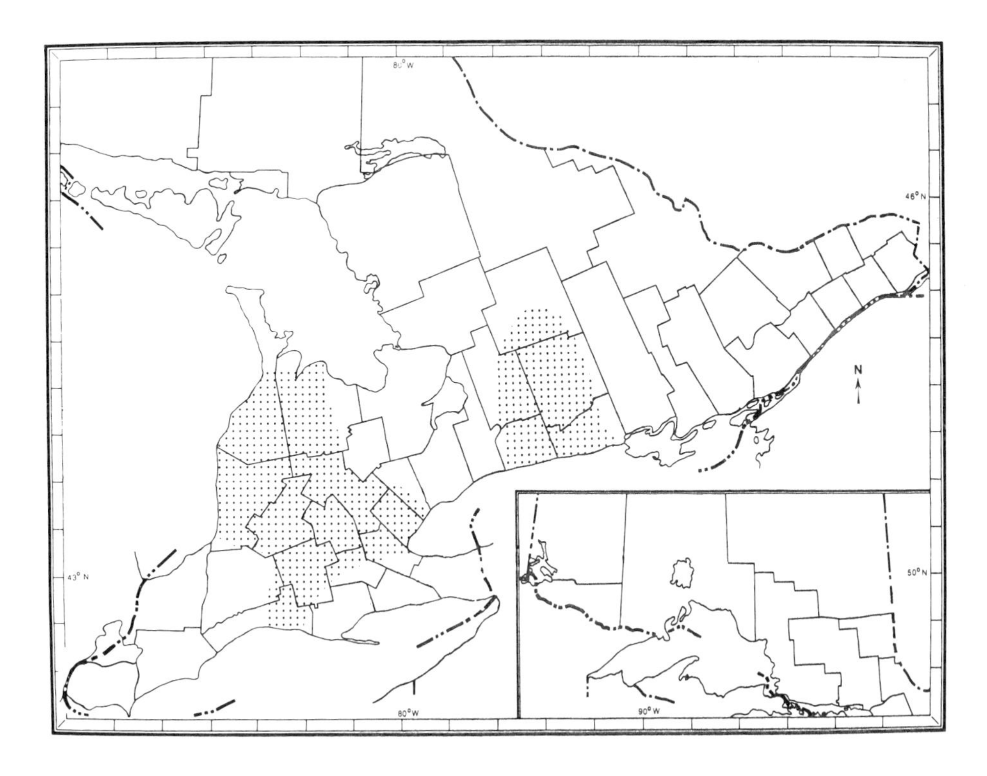

Fig. 39 The known Ontario distribution of *Octolasion tyrtaeum*.

LAND CO. *Hwy 45, 4.84 km n of Baltimore, under logs, 28 Apr 72, JWR, 1-4-1. OXFORD CO. *Hwy 97, Washington, w.e., under logs, 3 May 72, JWR, 0-1-3. PEEL CO. *Hwy 5, .81 km e of Dixie Rd, under logs, 29 Apr 72, JWR, 0-1-1. PETERBOROUGH CO. *Hwy 28, Lakefield College, under logs near waterfront, 27 Apr 72, JWR & CWR, 4-5-8. VICTORIA CO. *Hwy 7, 7.26 km e of Hwy 35, under log, 26 Apr 72, JWR, 0-3-1. WATERLOO CO. *Hwy 24A, 11.45 km n of Paris, under log, 3 May 72, JWR, 4-0-2. *Hwy 97, 3.71 km w of Hwy 401, under log, 3 May 72, JWR, 0-1-0. Hwy 7-8, 1.77 km w of Petersburg, under log, 3 May 72, JWR, 1-2-1. Waterloo, Beechwood South, on sidewalk after rain (evening), 15 Jun 75, 0-0-1, DPS, UW-0004. WELLINGTON CO. *Hwy 6, 2.26 km n of Aberfoyle, under log, 29 Apr 72, JWR, 0-0-1. *Hwy 24, 5.16 km e of Eramosa, under logs in cedar (*Thuja occidentalis*) woodlot, 29 Apr 72, JWR, 0-1-1. *Hwy 6, 10.65 km s of Arthur, under log, 2 May 72, JWR, 0-0-1. WENTWORTH CO. Reynolds (1972a). *Hwy 6, 5.32 km n of Hwy 5, under logs, 29 Apr 72, JWR, 1-0-2.

Family SPARGANOPHILIDAE Michaelsen, 1921

Diagnosis

Digestive system: without gizzard, calciferous glands, lamellae, caeca, typhlosole, or supra-intestinal glands, with an intestinal origin in front of the testis segments. Vascular system: with complete dorsal and ventral trunks, two pairs of anterior lateroparietal trunks, one of which passes to the dorsal vessel and the other to the ventral vessel in xiv, but without subneural and supra-oesophageal trunks. Hearts: lateral, moniliform, in vii–xi. Blood glands: protuberances from capillaries in septal glands. Nephridia: holoic, aborted at maturity in first 12 or more segments, avesiculate, peritoneal cells investing postseptal portions enlarged, ducts without muscular thickening passing into parieties in *AB*. Nephropores: inconspicuous, in *AB*. Setae: eight per segment. Dorsal pores and pigment, lacking. Prostomium, zygolobous. Anus, dorsal. Reproductive apertures: all minute and superficial, female pores in xiv, spermathecal pores in front of testis segments. Clitellum multilayered, including male pore segment. Seminal vesicles, trabeculate. Ovaries, in xiii, each terminating in a single eggstring. Ova, not yolky. Ovisacs, in xiv, small and lobed. Spermathecae, adiverticulate (after Gates, 1972c: 314).

Type Genus

Sparganophilus Benham, 1892 by monotypy (Michaelsen, 1921).

Discussion

This family is still monotypic and contains but a few nearctic species, only one of which occurs in Canada. Originally, Benham placed the genus in the Rhinodrilidae, a taxon no longer employed by oligochaetologists. Following Michaelsen (1900b), most authors have placed *Sparganophilus* in various subdivisions of the Glossoscolecidae. In a recent review Brinkhurst and Jamieson (1971) still considered the genus as belonging to the Glossoscolecidae.

Genus *Sparganophilus* (Benham, 1892)

1892 *Sparganophilus* Benham, Quart. J. Micros. Soc. (n.s.), 34: 155.
1895 *Sparganophilus*–Smith, Bull. Ill. St. Lab. Nat. Hist. 4(5): 142.
1900 *Sparganophilus*–Michaelsen, Das Tierreich, Oligochaeta 10: 463.
1921 *Sparganophilus*–Michaelsen, Arch. Naturg. 86(A): 141.
1930 *Sparganophilus*–Stephenson, Oligochaeta (Oxford), p. 899.
1971 *Sparganophilus*–Brinkhurst and Jamieson, Aquatic Oligochaeta World, p. 810.
1972 *Sparganophilus*–Gates, Trans. Amer. Philos. Soc. 62(7): 314.

Type Species
Sparganophilus tamesis Benham, 1892.

Diagnosis
Calciferous gland and gizzard, absent. Hearts, in vii–xi. Nephridia, absent in cephalic region (segments i–xii). Nephropores, inconspicuous, in *AB*. Setae, paired. Prostomium zygolobic. Lateral lines, absent. Colour, unpigmented.

Sparganophilus eiseni Smith, 1895

American mud worm Ver américain de la vase
(Fig. 40)

1895 *Sparganophilus eiseni* Smith, Bull. Ill. St. Lab. Nat. Hist. 4(5): 142.
1906 *Sparganophilus eiseni*–Moore, Bull. Bur. Fish. 25: 153.
1911 *Helodrilus elongatus* Friend, Zoologist, ser. 4, 15: 192.
1921 *Sparganophilus elongatus*–Friend, Ann. Mag. Nat. Hist., ser. 9, 7: 137.
1934 *Pelodrilus cuenoti* Tétry, C.R. Acad. Sci. Paris 199: 322.
1934 *Eiseniella tetrahedra* (laps.)–Moon, J. Anim. Ecol. 3: 17.

Diagnosis
Length 150–200 mm, diameter uniformly about 2.5 mm, segment number 165–225, prostomium zygolobic, dorsal pores absent. Clitellum xv–xxv. Tubercula pubertatis xvii–xxii. Setae closely paired, with setae *c* and *d* above mL line. Male pores on xix, female pores on xiv. Seminal vesicles, two pairs, in 11 and 12 with a pair of testes in 10 and 11. Spermathecae, three pairs, in 7, 8, and 9 with ducts opening just ventral to level *C* in 6/7–8/9. Gizzard, calciferous glands, and typhlosole absent. Prostate-like glands, four pairs, in 13, 14, 15, and 16. Body cylindrical. Colour, unpigmented but sometimes appearing pink with blue and green iridescence.

Biology
This species is limicolous and confined to the muddy banks of streams, rivers, ponds, and lakes. Smith (1915) found it to be abundant in the mud of the bottom and margins of many waters east of the Mississippi River. Olson (1928) reported it along the shores of Lake Erie, and similar habitats are recorded for New York (Lake Ontario) (Olson, 1940), New Jersey (Davies, 1954), Louisiana (Harman, 1965) and Tennessee (Reynolds, MS).

S. eiseni is relatively rare in North America except for an area bounded by the Great Lakes, Mississippi River, Atlantic Ocean, and the Gulf States. It is assumed to be amphimictic (Reynolds, 1974c).

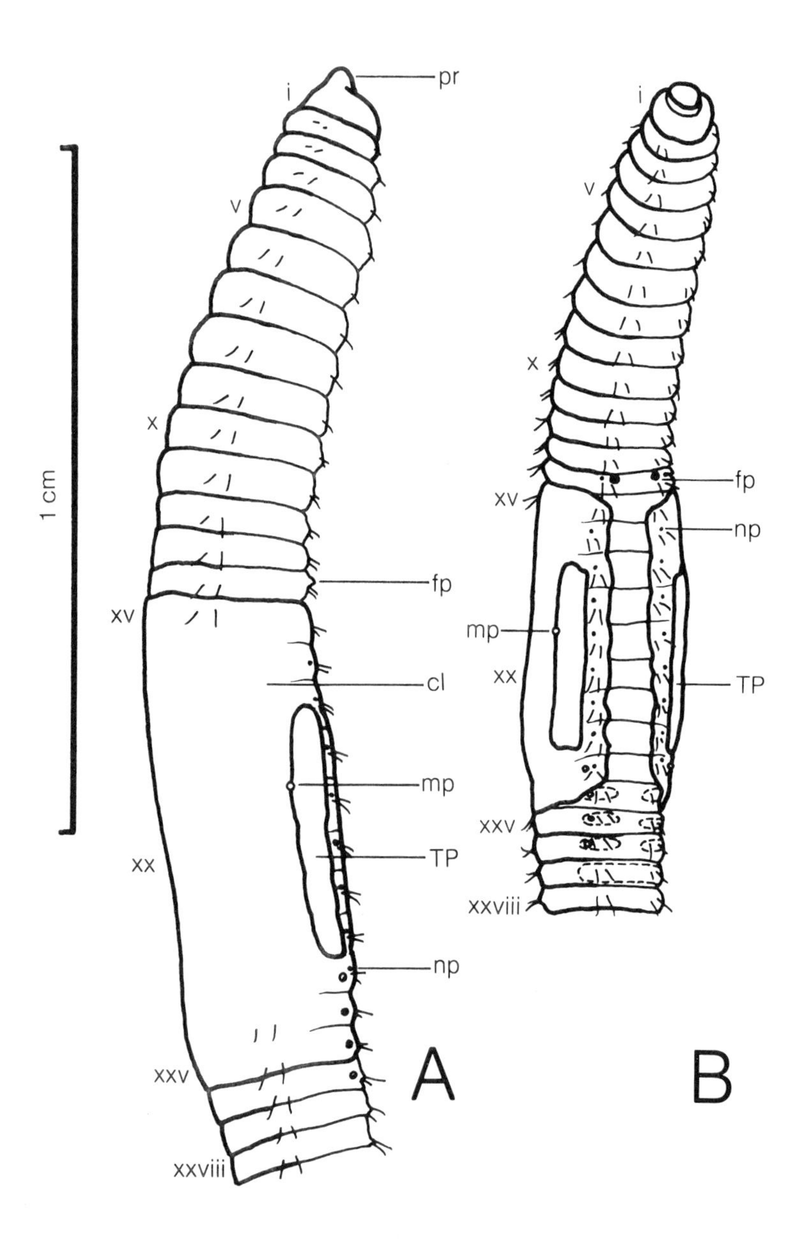

Fig. 40 External longitudinal views of *Sparganophilus eiseni* showing taxonomic characters. A. Dorsolateral view. B. Ventrolateral view. (GA: Thomas Co., cat. no. 3285)

Range

A native of North America, *S. eiseni* is known also from Central America and has been introduced into England and France (Brinkhurst and Jamieson, 1971; Gates, 1972c).

North American Distribution

Ontario (Moore, 1906), Alabama (Gates, 1967), Connecticut (Reynolds, 1973c), Florida (Smith, 1896), Illinois (Smith, 1895), Indiana (Heimburger, 1915), Iowa (Hague, 1923), Louisiana (Harman, 1965), Maryland (Reynolds, 1974b), Massachusetts (Reynolds, 1977), Michigan (Moore, 1906), New Jersey (Davies, 1954), New York (Olson, 1940), Ohio (Olson, 1928), Oregon (MacNab and McKey-Fender, 1947). New records: Arkansas, Georgia, Kentucky, Mississippi, North Carolina, Pennsylvania, Tennessee, Virginia, Wisconsin.

Ontario Distribution (Fig. 41)

Sparganophilus eiseni has been previously recorded from only two sites in Ontario (Moore, 1906).

BRUCE CO. Howdenvale, 2 metres depth, Lake Huron, airlift sampler, 11 Jul 75, DRB, 0-1-0, UW-0001. HALDIMAND CO. Selkirk Provincial Park, 2 metres depth, Lake Erie, airlift sampler, 5 Aug 75, DRB, 0-1-3, UW-0001. HALTON CO. Sixteen Mile Creek, intersection of creek and Lower Baseline Road, 1 Jun 71, AT, 0-0-6, ROM. KENT CO. Rondeau Harbour, 30 Aug 99, JPM, 1-1-0, ANSP-418a (Moore, 1906). MANITOULIN DIST. 2 metres depth, Georgian Bay, airlift sampler while scuba diving, 17 Aug 74, DRB, 0-0-1, UW-0001. NORFOLK CO. Long Point, 23 Aug 99, JPM, 4-0-0, ANSP-382 (Moore, 1906). PARRY SOUND DIST. Parry Sound, 3 metres depth, in sheltered shallow bay, Ekmann dredge, 18 Aug 75, RLH, 0-0-1, UW-box 1.

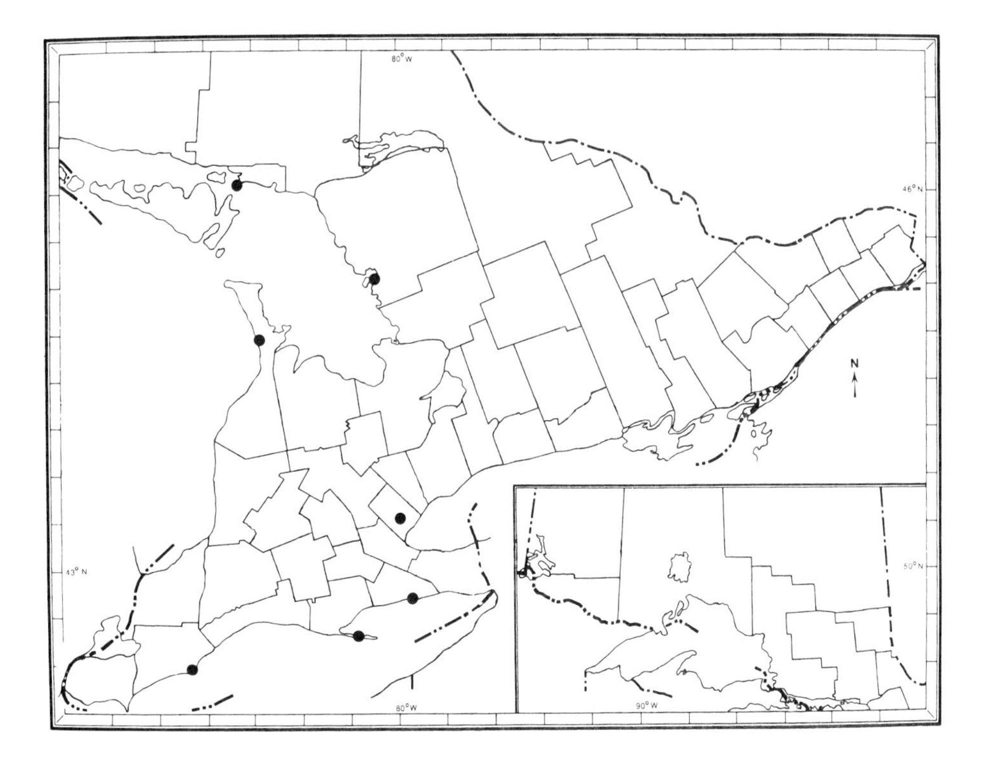

Fig. 41 The known Ontario distribution of *Sparganophilus eiseni*.

Distribution and Ecology

Even among oligochaetologists the question of megadrile distribution, particularly in North America, has long been controversial. Except for the southern tip of Queen Charlotte Island (British Columbia), almost the entire megadrile fauna in Canada is composed of species of the family Lumbricidae that are primarily European forms. Of the 19 Ontario species, for example, 18 are lumbricids. Two species, *Bimastos parvus* (Lumbricidae) and *Sparganophilus eiseni* (Sparganophilidae), are natives of North America.

There are 16 currently accepted genera in the family Lumbricidae: *Allolobophora, Aporrectodea, Bimastos, Dendrobaena, Dendrodrilus, Eisenia, Eiseniella, Eiseniona, Eisenoides, Eophila, Helodrilus, Kritodrilus, Lumbricus, Octolasion, Octodrilus,* and *Satchellius*. Of these, the genera *Eiseniona, Eophila, Helodrilus, Kritodrilus, Octodrilus*, and *Satchellius* are restricted to Europe and a few adjacent areas, such as North Africa (in the case of *Eophila*), except that four specimens of *Satchellius* were recorded in New Jersey. Two genera, *Bimastos* and *Eisenoides*, are nearctic but are restricted to the southeastern region of the United States except for two species, *Bimastos beddardi* and *B. parvus*, which have been carried around the world. *B. parvus* has been reported from Canada (Reynolds, 1972a), but subsequent investigation (E.L. Bousfield, pers. comm., 1973) nullified this as the only Canadian natural record because the one collection of four clitellate adults came from the Arboretum at the Central Experimental Farm, Department of Agriculture, Ottawa.

There are five other nearctic megadrile genera: *Argilophilus* and *Diplocardia* (Acanthodrilidae), *Komarekiona* (Komarekionidae), *Lutodrilus* (Lutodrilidae), and *Sparganophilus* (Sparganophilidae) (Reynolds, 1976b; Gates, 1977, 1974d, McMahan, 1976). The first four have not been reported from Canada or from any areas north of the southern limit of Quaternary glaciation in the United States. *Sparganophilus eiseni* has been reported from Canada once (Moore, 1906) but has never been recorded again in Canada until now. This species has also been introduced, probably by man, into Britain and France.

There is little disagreement among specialists that the centre of origin of the Lumbricidae is in Europe. Here the family is a very diverse group and the number and variety of species in the genera are greater than anywhere else in the world. There is less agreement, however, as to how those lumbricids became established in North America, a problem that has been discussed in detail most recently by Gates (1967, 1970) and Omodeo (1963) and reviewed by Reynolds et al. (1974) and Reynolds (1975f, 1976f).

Assuming that lumbricids have indeed travelled from Europe to North America, either they must have migrated actively by their own power, or they must have been transported passively by some agent. Since the Atlantic Ocean now forms an effective barrier to their active migration some workers have hypothesized that long ago they arrived in North America by migration via a landbridge, or were perhaps aided by continental drift (e.g., Omodeo, 1963). Such explanations, however, clearly are inadequate to explain the occurrence of some of the same species in South Africa, Australasia, and South America. Moreover, if such hypotheses are accepted then it is necessary to explain why it should be that of all the North American earthworms only the European lumbricids survived the Quaternary glaciations that otherwise depauperated the North American megadrile fauna.

In fact Gates (1961, 1970) and Reynolds (1975f, 1976f) have assembled a great deal of data indicating that during the Quaternary all the earthworms of glaciated North America were exterminated and such extermination extended, perhaps, almost as far as the area north of the Appalachian Mountains. Early settlers, for example, found that glaciated portions of the United States and Canada had no earthworms excepting where they were introduced (see Gates, 1967). Consequently, and contrary to the usual belief, surviving species did not follow the retreating glaciers. The earthworm species currently found in regions formerly covered by the ice sheets are all introductions, probably by man, either from other continents or from refugia south of the Appalachians. In recording the occurrence of some European lumbricids in South Africa, Australasia, and South America, the efficacy of transportation by man has rarely been questioned and there is no reason why this method of introduction need be denied for the large European element in the North American earthworm fauna. Gates (1966) therefore suggested that the colonization of North America by European lumbricids has occurred since 1500 A.D. and was made primarily by European settlers who brought potted plants and other agricultural material with them to the New World. Certainly this hypothesis explains both the diversity of the European element in glaciated North America and the near absence of nearctic endemics in the same region. The relative merits of the opposing views of Omodeo and Gates have recently been assessed from a theoretical viewpoint by Ball (1975), who concluded that the rival hypotheses were not mutually exclusive, and that on logical grounds it was not possible to prefer one over the other.

Our detailed knowledge of megadrile distribution within North America is rather limited and a summary of the North American biogeographical surveys to date is presented in Table 1.

Little work has been done on the ecology and distribution of Canadian earthworms and it is an open field of study. From a study of 14 species from Maine, 13 of which occur also in Ontario, Gates (1961) concluded that dietary preferences might be an important factor influencing distribution. He divided these species into three groups.

The first group comprised those species that pass much soil through the intestine and hence are termed geophagous, viz., *A. chlorotica, Ap. longa, Ap. tuberculata, Ap. turgida, E. rosea, L. terrestris*, and *O. cyaneum*. These species are known from soils of pH as low as 4.5–5.5 and the kinds of soil seem to be of little significance. Within this group only *L. terrestris* normally and regularly copulates at the surface. Because these species are tolerant of different soil types and occur in a wide range of habitats they are likely to be widely distributed, aided especially by the activities of man.

The second group, containing only *El. tetraedra*, is limiphagous (mud-eating) or limicolous (mud-inhabiting). Outside of Maine other species such as *Sparganophilus eiseni* and possibly *A. chlorotica* belong in this group, the members of which thrive in mud well under water or in saturated soils. The fact that on a few occasions *El. tetraedra* has been found to be geophagous suggests an adaptability that would prove advantageous if the animal were introduced to a new area.

The third group consisted of the litter feeders: *E. foetida, D. octaedra, Dd. ru-*

Table 1. Regional earthworm surveys in North America.

Region	Number of species obtained	Number of counties sampled	Total number of counties in state or province	References
Arkansas	21	22	75	Causey (1952, 1953)
Cape Breton	14	4	4	Reynolds (1975a)
Connecticut	21	8	8	Reynolds (1973c)
Delaware	15	3	3	Reynolds (1973a)
Gaspé	15	8	8	Reynolds (1975e)
Îles-de-la-Madeleine	9	8	10[1]	Reynolds (1975b)
Indiana	25+	92	92	Reynolds (MS)
Louisiana	8[2]	64	64	Tandy (1969)
Maryland	26	23	23	Reynolds (1974b)
Massachusetts	21	13	14	Reynolds (1976e)
Michigan	18	51	83	Murchie (1956)
Missouri	21	31	115	Olson (1936)
New Brunswick	13	15	15	Reynolds (1976d)
New York	18	29	62	Olson (1940), Eaton (1942)
Nova Scotia	15	18	18	Reynolds (1976a)
Ohio	22	55	88	Olson (1928, 1932)
Ontario	18	52	54	Reynolds[3]
Oregon	25	10	36	MacNab and McKey-Fender (1947)
Prince Edward Island	12	3	3	Reynolds (1975c)
Québec (sud côte)	15	37	37	Reynolds (1975d, 1976c)
Tennessee	44+[4]	95	95	Reynolds (1974a, MSS)
Washington	16	9	39	Altman (1936)

[1]Municipalités and îles.
[2]Tandy was concerned only with the genus *Pheretima*. Other reports by Gates (1965 and 1967) bring the state list to 25 species.
[3]Data included in this book.
[4]Includes species currently in manuscript.

bidus, *L. castaneus*, and *L. rubellus*. Litter includes all types of accumulation of organic matter such as leaves, manure, compost, etc. The activities of farmers and horticulturists greatly affect the distribution of these forms, and also, of course, of geophagous species that may stray into organic matter. In contrast, prior to hibernation, litter feeders become geophagous when they burrow down to the levels where they hibernate.

The extent to which Julin's (1949) ecological classification of the Lumbricidae (p. 5) can be applied to the Ontario earthworms is unknown. Only Reynolds et al. (1974) have attempted to apply the system to a North American fauna. Considering here only those species common to both Tennessee and Ontario, they placed *Allolobophora chlorotica*, *Dendrobaena octaedra*, *Dendrodrilus rubidus*, *Eisenia foetida*, *E. rosea*, *Eiseniella tetraedra*, *Lumbricus rubellus*, and *Octolasion cyaneum* among the hemerophiles. The hemerodiaphores included *Ap. trapezoides*, *Ap. turgida*, and *O. tyrtaeum*, and information was lacking for *Ap. longa*, *Bimastos parvus*, and *Lumbricus terrestris*. There were none that appeared to be true hemerophobes.

There is evidence to suggest, however, that this classification is not strictly applicable to Ontario. Thus, in Tennessee *Ap. tuberculata, E. rosea*, and *L. terrestris* are sparsely distributed and probably will become widespread and established only with the aid of human culture, whereas in Ontario these species already are widespread in great numbers (Reynolds et al., 1974). For these species there is a change from the hemerophilic to the hemerodiaphoric category from Tennessee to Ontario. The reverse is the case for *Ap. trapezoides* and *Octolasion tyrtaeum*, for these are of restricted distribution in Ontario and widespread in Tennessee (Reynolds et al., op. cit.).

There is no cause to consider any of the local lumbricids as hemerobionts except in areas where the natural habitats have been completely destroyed, as in the Sudbury–Copper Cliff region of the Sudbury District of Ontario.

Ontario lumbricids, like those of Maine (Gates, 1961), must have become habituated to rather low temperatures. Nothing is known about the state in which winter is spent but, in Maine at least, *E. rosea* has been found active under leaves surrounded by snow and frost and active specimens of *Ap. tuberculata, L. castaneus, D. octaedra* and *Dd. rubidus*, some in breeding condition, have been found in a ridge surrounded with 75-cm deep snow. *L. terrestris* has been found at night on frozen ground and it seems that melt water from snow and ice is not too cold for many of these lumbricids. Presumably the same holds for the Ontario species, though it is evident that a study of environmental and climatic factors affecting the distribution and activity of the Ontario fauna is greatly needed.

In determining the presence and distribution of the earthworms in Ontario, 330 new county and district records were established. Ten of these new distribution records were found in the ROM unidentified collections by the author, two came from Carleton University collections, and the remainder were collected by the author. For a detailed breakdown of these records see Table 2 and the Ontario Distribution records for each species.

When careful and thorough collection procedures are followed, earthworm species will be found in associations. The size and number of associations obtained in this study are reported in Table 3. The maximum number of species found at any one site was eight, while single species, or associations of two or three, were most frequently encountered.

Associations vary considerably in size depending on habitat and geography. Reviews of some previously reported habitat associations were presented by Bouché (1969b) and by Reynolds (1972b, 1973d). Since then, another study from one of Bouché's (1969b) stations has been reported (Reynolds, 1972d). This latter report records lumbricid species associations of two, five, and eight for woodlots in north central France. In the previous Ontario study (Reynolds, 1972a), species associations ranged from four with most sites in the single species and two species association categories. The most common association was *Aporrectodea tuberculata—Lumbricus rubellus*. Recent lumbricid studies in Tennessee (Reynolds et al., 1974) involving almost 1,700 sites and 18,000 specimens, recorded associations most frequently of between three and four species per site with a range of one to eight species. When the remaining families of earthworms have been analyzed these values should increase.

The most frequent associations found in this study were: 1) *Aporrectodea tuberculata—Aporrectodea turgida—Eisenia rosea*, 2) *Aporrectodea*

Table 2. Summary of numbers of new county Distribution Records for Ontario earthworms recorded herein.

Species	Number of new county records	First provincial record
Allolobophora chlorotica	25	Stafford (1902)[1]
Aporrectodea icterica	0	Schwert (1977)
Aporrectodea longa	0	Smith (1917)
Aporrectodea trapezoides	34	Reynolds (1972a)
Aporrectodea tuberculata	41	Eisen (1874)
Aporrectodea turgida	41	Eisen (1874)
Bimastos parvus	0	Reynolds (1972a)
Dendrobaena octaedra	9	Reynolds (1972a)
Dendrodrilus rubidus	25	Eisen (1874)
Eisenia foetida	7	Stafford (1902)[1]
Eisenia rosea	39	Stafford (1902)[1]
Eiseniella tetraedra	11	Eisen (1874)
Lumbricus castaneus	1	Eisen (1874)
Lumbricus festivus	7	Stafford (1902)[1]
Lumbricus rubellus	33	Stafford (1902)[1]
Lumbricus terrestris	37	Stafford (1902)[1]
Octolasion cyaneum	1	Reynolds (1972a)
Octolasion tyrtaeum	14	Reynolds (1972a)
Sparganophilus eiseni	5	Moore (1906)
Total	330	

[1]First reported by Stafford (1902) but he gave no collection details and no locations other than Ontario, Québec, New Brunswick, and Nova Scotia.

tuberculata—Aporrectodea turgida—Lumbricus rubellus, 3) *Aporrectodea tuberculata—Lumbricus terrestris*, 4) *Aporrectodea tuberculata—Aporrectodea turgida*, and 5) *Aporrectodea tuberculata* or *Aporrectodea turgida—Dendrodrilus rubidus*. Because of the limited diversity of earthworms in Ontario, one should not be surprised at the repetitive nature of the species associations. Even though 19 species have been recorded from Ontario, this does not represent a particularly diverse fauna when compared with other political regions of North America (Table 1) because the fauna comprises principally exotic European forms; the nearctic endemic species characteristic of more southern regions are absent. Tennessee, for example, has over 44 species known to exist in various habitats within its boundaries (Reynolds et al., 1974). Three of Ontario's 19 species (*Apporectodea icterica, Lumbricus castaneus*, and *L. festivus*) are not found in Tennessee, the remainder are. The many species in Tennessee not found in Ontario belong mainly to the nearctic genera *Bimastos, Diplocardia, Eisenoides*, and *Komarekiona*. Recent observations by this author (Reynolds, 1973a, 1973b, 1973c, and 1974b) indicate that the habitats utilized by the nearctic species in the southern portions of the United States are those which in Ontario are invaded and utilized by exotic European lumbricids. The success of these species in these habitats no doubt is a result of the lack of competition from endemic forms.

Table 3. Earthworm associations in Ontario by counties and districts.

Counties and districts	Number of collections in county or district	Frequency of obtaining the following number of species per collection							
		1	2	3	4	5	6	7	8
Algoma	3	0	.33	.33	.33	0	0	0	0
Brant	6	0	.50	.33	0	.17	0	0	0
Bruce	8	.12	.12	.12	.38	.12	0	.12	0
Carleton	5	0	.20	.60	.20	0	0	0	0
Cochrane	3	0	.67	0	0	.33	0	0	0
Dufferin	7	0	.42	.29	0	0	0	0	0
Dundas	7	.14	.57	.29	.29	0	0	0	0
Durham	6	.33	0	.33	0	.17	.17	0	0
Elgin	8	.12	.25	.12	.25	.12	.12	0	0
Essex	7	.29	.29	.14	.29	0	0	0	0
Frontenac	8	.25	.25	.38	.12	0	0	0	0
Glengarry	4	0	.25	.75	0	0	0	0	0
Grenville	5	0	.40	.40	.20	0	0	0	0
Grey	6	0	0	.83	.17	0	0	0	0
Haldimand	6	.33	0	.17	.33	.17	0	0	0
Haliburton	72	.52	.33	.12	.03	0	0	0	0
Halton	9	.33	.44	0	.11	0	.11	0	0
Hastings	7	.43	.43	.14	0	0	0	0	0
Huron	7	0	0	.57	.29	.14	0	0	0
Kent	6	.33	0	.33	.17	0	.17	0	0
Kenora	1	0	1.00	0	0	0	0	0	0
Lambton	6	.17	.50	.33	0	0	0	0	0
Lanark	5	.20	.20	.40	.20	0	0	0	0
Leeds	7	0	.72	.14	0	.14	0	0	0
Lennox & Addington	5	0	.60	.40	0	0	0	0	0
Manitoulin	6	.17	.17	.17	.50	0	0	0	0
Middlesex	5	.20	.20	.20	.20	0	0	.20	0
Muskoka	9	.56	.44	0	0	0	0	0	0
Niagara	4	0	0	.25	.50	.25	0	0	0
Nipissing	6	.33	0	.50	0	.17	0	0	0
Norfolk	7	.57	.29	0	0	.14	0	0	0
Northumberland	9	.44	.11	.33	.11	0	0	0	0
Ontario	7	.43	.14	0	.29	0	.14	0	0
Oxford	7	.29	.14	.14	.29	.14	0	0	0
Parry Sound	11	.45	.45	.10	0	0	0	0	0
Peel	6	.17	0	.33	.33	.17	0	0	0
Perth	4	0	.50	0	0	.25	0	.25	0
Peterborough	4	.25	25	.25	.25	0	0	0	0
Prescott	5	.20	.60	.20	0	0	0	0	0
Prince Edward	7	.14	.29	.57	0	0	0	0	0
Renfrew	11	.73	.18	.09	0	0	0	0	0
Russell	4	0	.25	.25	.50	0	0	0	0
Simcoe	8	0	.12	.63	.12	.12	0	0	0
Stormont	5	.40	.40	0	.20	0	0	0	0
Sudbury	7	.72	.14	.14	0	0	0	0	0
Thunder Bay	1	1.00	0	0	0	0	0	0	0
Victoria	7	.14	.14	.43	.29	0	0	0	0
Waterloo	10	.10	0	.60	.30	0	0	0	0
Wellington	6	0	.17	.33	.17	.33	0	0	0
Wentworth	5	.20	0	.20	.20	0	0	.40	0
York	10	.30	.10	.20	.10	.10	0	.10	.10
Total collections	384								
Average frequency of obtaining the number of species per collection		.22	.26	.26	.15	.06	.01	.02	.002

Appendix: Provincial Description

It is the author's hope that this text will facilitate and stimulate further studies of North American earthworms. Certainly we need to know more concerning their detailed distribution if we are to assess fully their biological importance. For Ontario readers, therefore, a provincial description now follows. This is intended to form a basis upon which regional surveys can be made. It is to be hoped that ultimately we will be able to correlate distribution with edaphic and vegetational factors. The provincial description also will serve as a reference point for foreign readers unfamiliar with the Canadian environment.

Southern Ontario lies between 42° and 46° N latitude and 75° and 83° W longitude (Fig. 1, p. 2) in the St. Lawrence Drainage Basin and primarily in the Plains of the lower Great Lakes and St. Lawrence Lowlands, a physiographic region underlain by relatively undisturbed Paleozoic sedimentary beds of limestone, shale, and sandstone. The northern portion of southern Ontario (Algoma, Cochrane, Manitoulin, Nipissing, and Sudbury districts) is underlain by Precambrian formations in a physiographic region known as the Canadian Shield. All of Ontario has been glaciated.

Since 1915, soil surveyors have been at work in Canada, and most of the settled area has been mapped on a preliminary scale at least, except for the great unoccupied areas of northern Canada. One should consult Ehrlich (1968) and United States Department of Agriculture (1964, 1967 and 1968) for a detailed explanation of the soil classification schemes used in the following paragraphs and in other megadrile surveys.

Southern Ontario contains large expanses of three soil orders—Alfisols, Spodosols, and Inceptisols. The Alfisols (Aquic and Ochreptic Hapludalfs), mineral soils, also known as Grey Brown Podzolic soils, are located in two areas: 1) a line through Simcoe, Victoria, Peterborough, Northumberland, and Prince Edward counties and all counties to the west, and 2) a line through Carleton and Leeds counties and all counties to the east. The parent materials either are glacial in origin or were deposited in the great bodies of water which occupied the lowlands at the close of the glacial epoch. Under deciduous forests the leaching is not intense. The surface soil (A horizon) is greyish (10YR 4/2–5/2) (Munsell, 1954; USDA, 1951) and slightly acid, and has a moderate amount of organic matter. The B horizon is brown (10YR 5/3, 4/3, 3/2, 2/2) and nut-like in structure because of the accumulation of clay (argillic B horizon). Normally, all lime is leached from the soil profile which may be slightly acid throughout. Generally the soils of this suborder (Udalfs) are reasonably fertile and well suited to cultivation.

The Spodosols appear to be more weathered than the Alfisols. Spodosols are mineral soils with a spodic epipedon and a subsurface horizon with an accumulation of organic matter and aluminium oxides $\pm$ iron oxides. These soils (Spodosols: Boralfic Cryorthods, Typic Cryorthods, and Humic Haplorthods) are also known as Bisequa and Orthic Podzolic soils and occupy the major region not included above for the Alfisols. These soils form mostly on coarse-textured, acid parent material subject to ready leaching occurring in humid climates commonly where it is cold and temperate. Forests are natural cover vegetation. Conifers, low in metallic ion content, seem to encourage the development of Spodosols. As the litter from these low-base species decomposes, a strong acidity develops and percolating water leaches acids down into the profile. Since the upper horizons are so intensly leached, Spodosols are not naturally fertile. If properly fertilized, these soils can become quite productive.

The inceptisols are young mineral soils whose profiles contain horizons which 1) have formed quickly, 2) result mostly from alteration of parent materials, 3) are free of extreme weathering, and 4) lack accumulation of clay, and iron and aluminium oxides. These Degraded Brown Forest soils (Mollic Alfic Eutrochrepts) are located mainly on a line from Bruce to Glengarry counties. These soils are generally in forests as agricultural productivity may be limited without considerable expense.

These three soil groups make up nearly all of the soils of southern Ontario. But there are other minor areas of note, such as the sands or Orthic Regosols (Entisols: Udipsamments and Udorthents) in portions of Durham, Haliburton, Norfolk, and Victoria counties. Also there are many areas comprised of Histosols (Fibrists, Hemists, and Saprists), known previously as Bog and/or Half-bog soils. These organic or bog soils are found covering sizeable areas of Essex, Kent, Lambton, and Simcoe counties. In both cases, these groups (Entisols and Histosols) have in recent decades been converted into areas of productive specialized cropping.

Of the 14 major regions of vegetation in Canada, southern Ontario contains part of the Great Lakes–St. Lawrence Forest and all of the Niagara Forest (Rowe, 1972).

The Great Lakes–St. Lawrence Forest extends from Lake-of-the-Woods to Baie de Chaleur and is essentially a transition between the boreal coniferous forest and the deciduous forest of eastern North America. The dominant conifers are: white pine (*Pinus strobus* L.), red pine (*P. resinosa* Ait.), hemlock (*Tsuga canadensis* (L.) Carr.), and white cedar (*Thuja occidentalis* L.); others, apparently invaders from the north, are: jack pine (*Pinus banksiana* Lamb.), tamarack (*Larix laricina* (Du Roi) K. Koch), balsam fir (*Abies balsamea* (L.) Mill.) and white spruce (*Picea glauca* (Moench) Voss). The dominant hardwoods are: sugar maple (*Acer saccharum* Marsh.), beech (*Fagus grandifolia* Ehrh.), yellow birch (*Betula alleghaniensis* Britt.), red oak (*Quercus rubra* L.), bur oak (*Quercus macrocarpa* Michx.) and white oak (*Quercus alba* L.) on upland soils; with red maple (*Acer rubrum* L.), silver maple (*Acer saccharinum* L.), elm (*Ulmus*) and ash (*Fraxinus*) in the low ground. This forest region has probably more species and a greater number of associates than any other in Canada. The vegetation also includes many smaller plants, shrubs, and herbaceous forms on the forest floor and cleared lands. Some of the more prominent are: ground ivy (*Glecoma hederacea* L.), juniper (*Juniperus virginiana* L.), witch-hazel (*Hamamelis virginiana* L.), sumach (*Rhus typhina* L.), poison ivy (*Rhus radicans* L.), serviceberry (*Amelanchier oblongifolia* T. & G.), wild grape (*Vitis labrusca* L.), hawthorn (*Crataegus foetida* Ashe), raspberry and blackberry (*Rubus* spp.), thimbleberry (*Rubus odoratus* L.), and honeysuckle (*Lonicera tatarica* L.). Hawthorns have taken over many thousands of hectares of pasture land, and large areas are occupied by tangles of raspberries and brambles. There are many herbaceous plants on the floor of the deciduous forest, such as may apple (*Podophyllum peltatum* L.), herb Robert (*Geranium Robertianum* L.), Jack-in-the-pulpit (*Arisaema triphyllum* (L.) Schott), lily-of-the-valley (*Maianthemum canadense* Desf.), trillium (*Trillium grandiflorum* (Michx.) Salisb.), and sarsaparilla (*Aralia nudicaulis* L.). Many species of aster (*Aster* spp.) and goldenrod (*Solidago* spp.), almost unnoticed in the forest, dominate large areas of unimproved, low pasture land. Dry sites are covered by mullein (*Verbascum thapsus* L.) and blueweed (*Echium vulgare* L.). The roadside flora is just as characteristic as that of the forest.

The Niagara Forest is a strip along the northern shore of Lake Erie (Essex, Kent, Lambton, Elgin, Middlesex, Norfolk, Brant, Niagara, Oxford, and Wentworth counties). Much of what was said concerning the composition and appearance of the vegetation of the larger Great Lakes–St. Lawrence Forest also applies to the Niagara Forest. However, there are differences. Except for the pines on the sand plains there are few conifers while, on the other hand, there are additional species of deciduous trees such as: blue ash (*Fraxinus quadrangulata* Michx.), tulip poplar (*Liriodendron tulipifera* L.), sassafras (*Sassafras albidum* (Nutt.) Nees), magnolia (*Magnolia acuminata* L.), Kentucky coffee tree (*Gymnocladus dioica* (L.) K. Koch), sycamore (*Plantanus occidentalis* L.), black walnut (*Juglans nigra* L.), pawpaw (*Asimina triloba* (L.) Dunal), and others. This area is also the habitat of many small plants not found farther north.

There are large areas of cultivated land in southern Ontario. Prior to European settlement there were no forages and few natural meadows. The natural meadows were assigned to settlers on a livestock number basis. The lack of feed made it necessary to seed and cultivate meadows for wintering cattle. Most of the forages are of European origin. Recent values for areas of cultivated land in Ontario are: seeded pasturages 1.35 million hectares, rough pasture >1.30 million hectares, and small grains slightly >2 million hectares.

One of the most popular grasses in pasture mixtures is timothy (*Phleum pratense* L.), which is used on 75–80% of Ontario farms. Other prominent pasture species are: orchard grass (*Dactylis glomerata* L.), brome grass (*Bromus inermis* Leyss.), tall fescue (*Festuca elatior* L.), reed canary grass (*Phalaris arundinacea* L.), Kentucky bluegrass (*Poa pratense* L.), Canada bluegrass (*Poa compressa* L.), red fescue (*Festuca rubra* L.), perennial ryegrass (*Lolium perenne* L.), alfalfa (*Medicago sativa* L.), red clover (*Trifolium pratense* L.), white clover (*Trifolium repens* L.), and birdsfoot trefoil (*Lotus corniculatus* L.). Various combinations of these species are found in Ontario depending on the soil type, annual precipitation, physiographic position, and topography of the pasture, etc. For example, Canada bluegrass is well adapted to drier areas while Kentucky bluegrass can only be sown in moist areas.

The major small grains sown in Ontario are: oats (*Avena sativa* L.), barley (*Hordeum vulgare* L.), wheat (*Triticum* spp.), rye (*Secale cereale* L.), sorghum (*Sorghum vulgare* Pers.), and very limited areas of rice (*Oryza sativa* L.).

The climate of Ontario, according to the Köppen system (Pettersen, 1968), is in the cool snow-forest climatic type (Dfb) with warm summers and the absence of a dry season. The Holdridge system (Sawyer and Lindsay, 1964) places the province in the cool temperate moist forest bioclimatic formation. Utilizing the

Thornthwaite (1948) classification of climate for Ontario, the province is characterized in the following manner:

1. Moisture Regions—Climatic Type (moisture deficiency surplus index): B_1 Humid (20–40)—south and west of a line through Huron and Wentworth counties, southern Kenora and Rainy River districts. B_2 Humid (40–60)—the rest of southern Ontario.
2. Seasonal Variation of Effective Moisture: r (little or no water deficiency in any season)—all of southern Ontario.
3. Average Annual Thermal Efficiency—Thermal Efficiency Type (TE index, in inches): B'_1 or Mesothermal (22.44–28.05)—south and west of a line through Huron and Wentworth counties. C'_2 or Microthermal (16.83–22.44)—the rest of southern Ontario.
4. Summer Concentration of Thermal Efficiency—Thermal Efficiency Type (%): b'_2 (56.3–61.6)—south and west of a line through Huron and Wentworth counties. b'_1 (61.6–68.0)—the rest of southern Ontario.

The mean annual precipitation for southern Ontario ranges from 711 mm (Essex and Niagara counties and Manitoulin District) increasing eastward to 916 mm (Prescott County). The mean January air temperature is –1 to –7° C (maximum) and –12 and –18 °C (minimum) while the mean-July air temperature is 24 to 27 °C (maximum) and 10 to 16 °C (minimum). For further details, see Brown et al. (1968) or Thomas (1953).

Literature Cited

ALLEE, W. C., M. M. TORVICK, J. P. LAHR and P. L. HOLLISTER
1930 Influence of soil reactions on earthworms. –Physiol. Zool. 3: 164–200.

ALTMAN, L. C.
1936 Oligochaeta of Washington. –Univ. Wash. Publ. Biol. 4(1): 1–137.

ANDRÉ, F.
1963 Contribution à l'analyse expérimentale de la reproduction des Lombriciens. –Bull. Biol. Fr. Belg. 81: 1–101.

ANDREWS, E. A.
1895 Conjugation of the brandling. –Amer. Nat. 29: 1021–1027.

ATKINSON, R. J. C.
1957 Worms and weathering. –Antiquity 31: 219–233.

AVEL, M.
1959 Classe des Annélides, Oligochètes. *In* Grassé, P. P., ed. Traité de Zoologie. –Paris, 5(1): 224–470.

BACKLUND, H. O.
1949 Oligochaeta. 1. Lumbricidae. –Zool. Iceland II(20a): 1–15.

BAERMANN, G.
1917 Eine einfache Methode zur Auffindung von Ankylostomum-(Nematoden)-Larven in Erdproben. –Meded. Geneesk. Lab. Weltevr., pp. 41–47.

BAHL, K. N.
1947 Excretion in Oligochaeta. –Biol. Rev. 22: 109–147.

BAIRD, W.
1873 *Megascolex antarctica*, an earthworm from New Zealand. –Proc. Linn. Soc. Lond. 11: 96.

BAKER, W. L.
1946 DDT and earthworm populations. –J. Econ. Ent. 39(3): 404–405.

BALL, I. R.
1975 Nature and formulation of biogeographical hypotheses. –Syst. Zool. 24(4): 407–430.

BALL, R. C. and L. L. CURRY
1956 Culture and agricultural importance of earthworms. –Michigan State University, Agricultural Experiment Station, Circular Bulletin 222: 27 pp.

BEDDARD, F. E.
1891 The classification and distribution of earthworms. –Proc. R. Phy. Soc. Edinburgh 10: 235–290.
1895 A monograph of the order of Oligochaeta. –Oxford, xxi + 769 pp.

BENHAM, W. B.
1890 An attempt to classify earthworms. –Quart. J. Micros. Soc., n.s., 31(2): 201–315.
1892 A new English genus of aquatic Oligochaeta (*Sparganophilus*) belonging to the family Rhinodrilidae. –Quart. J. Micros. Soc., n.s., 34: 155–179.

BERKELEY, C.
1968 Records of earthworms from Vancouver, British Columbia. –J. Fish. Res. Bd. Can. 25(1): 205.

BHATTI, H. K.
1965 Earthworms of Swarthmore Pennsylvania and vicinity. –Proc. Penn. Acad. Sci. 39: 8–24.

BLAKE, I. H.
1927 A comparison of the animal communities of coniferous forests and deciduous forests. –Ill. Biol. Monogr. 10(4): 1–148.

BLANCHARD, E.

1849 Escoleideaños. –Hist. Fisica Politica Chile (Zool.) 3: 37–52.

BLUMENBACH, J. F.

1825 Haubach der Naturgeschichte. 11th ed. –Obttingan, xi + 668 + 2 pl.

BORNEBUSCH, C. H.

1930 The fauna of forest soil. –Forstl. Forsøgs. Danm. 11: 1–224.

BOUCHÉ, M. B.

1969 a Comparison critique de méthodes d'évaluation des populations de lumbricidés. –Pedobiologia 9: 26–34.

1969 b La biogeographie des Lumbricidés de France, son intérêt et ses ambiguités. Cas d'*Allolobophora cupulifera* Tétry, *A. icterica* (Sav.), de *Lumbricus friendi* Cognetti et de *Lumbricus herculeus* (Sav.). –Pedobiologia 9: 87–92.

1970 Observations sur les Lombricidés (3ème séries: VII, VIII, IX). –Rev. Écol. Biol. Sol 7(4): 533–547.

1972 Lombriciens de France, Écologie et Systématique. –Paris, Inst. Natn. Rech. Agron., 671 pp.

1973 Commentaires: *Lumbricus terrestris* Linnaeus, 1758. –Bull. Zool. Nomencl. 30(2): 68.

BOUCHÉ, M. B. and M. BEUGNOT

1972 a Contribution à l'approche méthodologique de l'étude des biocenoses. II. L'extraction des macroéléments du sol par lavage-tamisage. –Ann. Zool. Écol. Anim. 4(4): 537–544.

1972 b La complexité de *Lumbricus herculeus* illustrée par les caractéristiques des populations de stations de la R. C. P. 40. –Rev. Écol. Biol. Sol 9(4): 697–704.

BRINKHURST, R. O. and B. G. M. JAMIESON

1971 Aquatic Oligochaeta of the world. –Toronto, University of Toronto Press, xi + 860 pp.

BROWN, D. M.

1944 Cause of death in submerged earthworms. –J. Tenn. Acad. Sci. 19: 147–149.

BROWN, D. M., G. A. MCKAY and L. J. CHAPMAN

1968 The climate of southern Ontario. –Can. Dept. Transp., Met. Br., Clim. Stud., no. 5, 50 pp.

BURMEISTER, H.

1835 Zoologischer Hand-Atlas zum Schulgebrauch und Selbstunterricht. –Berlin.

CAMERON, M. L. and W. H. FOGAL

1963 The development and structure of the acrosome in the sperm of *Lumbricus terrestris* L. –Can. J. Zool. 41: 753–761.

CAUSEY, D.

1952 The earthworms of Arkansas. –Proc. Ark. Acad. Sci. 5: 31–42.

1953 Additional records of Arkansas earthworms. –Proc. Ark. Acad. Sci. 6: 47–48.

ČERNOSVITOV, L.

1930 Studien über die Spermaresorption. I Tiel. Die Samenresorption bei den Oligochäten. –Zool. Jb. Anat. 52: 487–538.

1942 Oligochaeta from various parts of the world. –Proc. Zool. Soc. London, (B), 111: 197–236.

ČERNOSVITOV, L. and A. C. EVANS

1947 Lumbricidae. Synopses of the British fauna. 6. –Linn. Soc. London, 36 pp.

CHANDEBOIS, R.

1957 Une nouvelle forme, provençale et corse, de *Lumbricus castaneus* (Sav.). –Bull. Soc. Zool. France 82: 5–6, 417–419.

1958 *Dendrobaena rivulicola* n. sp., nouveau lumbricide amphibie de la région méditerranéenne. –Bull. Soc. Zool. France 83: 2–3, 159–162.

CHEN, Y.

1931 On the terrestrial Oligochaeta from Szechuan, with descriptions of some new forms. –Cont. Biol. Lab. Sci. Soc. China (Zool.) 7(3): 117–171.

CLAUS, C.
1876 Grundzüge der Zoologie. 3rd ed. Vol. 1. Aulf. -Marburg and Leipzig.
1880 Grundzüge der Zoologie. 4th ed. Vol. 1. -Marburg, N. G. Elwert'sche Univ.

CROSSLEY, D. A. JR., D. E. REICHLE and C. A. EDWARDS
1971 Intake and turnover of radioactive cesium by earthworms (Lumbricidae). -Pedobiologia 11: 71-76.

DARWIN, C. R.
1881 The formation of vegetable mould through the actions of worms, with observations on their habits. -New York, Appleton. vii + 326 pp.

DAVIES, H.
1954 A preliminary list of the earthworms of northern New Jersey with notes. -Breviora, Mus. Comp. Zool. no. 26, 13 pp.

DINDAL, D. L.
1970 Feeding behavior of a terrestrial turbellarian *Bipalium adventitium*. -Amer. Midl. Nat. 83(2): 635-637.

DOANE, C. C.
1962 Effects of certain insecticides on earthworms. -J. Econ. Ent. 55: 416-418.

DOBSON, R. M. and J. R. LOFTY
1956 Observations of the effect of BHC on the soil fauna of arable land. Congr. Int. Sci. Sol, Paris 3: 203-205.

DOEKSEN, J.
1950 An electrical method of sampling soil for earthworms. -Trans. 4th Int. Congr. Soil Sci., pp. 129-131.

DUGÈS, A.
1828 Recherche sur la circulation, la respiration et la reproduction des Annélides sétigères abranches. -Ann. Sci. Nat. 15(1): 284-336.
1837 Nouvelles observations sur la zoologie et l'anatomie des Annélides abranches sétigères. -Ann. Sci. Nat., ser. 2, Zool. 8: 15-35.

EATON, T. H. JR.
1942 Earthworms of the northeastern United States: a key, with distribution records. -J. Wash. Acad. Sci. 32(8): 242-249.

EDWARDS, C. A.
1965 Effects of pesticide residues on soil invertebrates and plants. Oxford, 5th Symp. Br. Ecol. Soc., pp. 239-261.
1970 Effects of herbicides on the soil fauna. -Proc. 10th Weed Control Conf. 3: 1052-1062.

EDWARDS, C. A. and E. B. DENNIS
1960 Some effects of aldrin and DDT on soil fauna or arable land. -Nature, Lond. 188(4572): 767.

EDWARDS, C. A. and J. R. LOFTY
1972 Biology of earthworms. -London, Chapman and Hall. xv + 283 pp.

EDWARDS, C. A., E. B. DENNIS and D. W. EMPSON
1967 Pesticides and the soil fauna. 1. Effects of aldrin and DDT in an arable field. -Ann. Appl. Biol. 60: 11-22.

EDWARDS, C. A., D. E. REICHLE and D. A. CROSSLEY, JR.
1969 Experimental manipulation of soil invertebrate populations for trophic studies. -Ecology 50(3): 495-498.

EDWARDS, C. A., A. E., WHITING and G. W. HEATH
1970 A mechanized washing method for separation of invertebrates from soil. -Pedobiologia 18(5): 141-148.

EDWARDS, R. M.
1967 Worms rise to a beat.-The Field 6(15): 2.

EHRLICH, W. A., ED.
1968 Outline of the Canadian system of soil classification for the order, great group and subgroup categories. –Edmonton, Proc. 7th Meeting Natn. Soil. Surv. Comm. Can. 216 pp.

EISEN, G.
1871 Bidrag till Skandinaviens Oligochaet-fauna. –Öfv. Vet.-Akad. Förh. Stockholm 27(10): 953–971.
1872 Om nagra arkiska Oligochaeter. –Öfv. Vet.-Akad. Förh. Stockholm 29(1): 119–124.
1873 Om Skandinaviens Lumbricider. –Öfv. Vet.-Akad. Förh. Stockholm 30(8): 43–56.
1874 New Englands och Canadas Lumbricider. –Öfv. Vet.-Akad. Förh. Stockholm 31(2): 41–49.

EVANS, A. C.
1946 A new earthworm of the genus *Allolobophora*. –Ann. Mag. Nat. Hist., ser. 11, 13: 98–101.
1948 a On some earthworms from Iowa, including a description of a new species. –Ann. Mag. Nat. Hist., ser. 11, 14: 514–516.
1948 b Identity of earthworms stored by moles. –Proc. Zool. Soc. Lond. 118: 1356–1359.

EVANS, A. C. and W. J. MCL. GUILD
1947 Studies on the relationships between earthworms and soil fertility. I. Biological studies in the field. –Ann. Appl. Biol. 34(3): 307–330.
1948 Studies on the relationships between earthworms and soil fertility. IV. On the life cycles of some British Lumbricidae. –Ann. Appl. Biol. 35(4): 471–484.

FABRICIUS, O.
1780 Fauna Grøenlandica. –Havniae et Lipsiae, pp. 321–323.

FOX, C. J. S.
1964 The effects of five herbicides on the number of certain invertebrate animals in grassland soils. –Can. J. Plant Sci. 44: 405–409.

FRIEND, H.
1891 The identification of Templeton's British earthworms. –Nature 44: 273.
1892 a British Annelida. –Essex Nat. 6: 30–33, 60–65, 107–111, 169–174, 185–190.
1892 b On a new species of earth-worm. –Proc. R. Irish Acad., ser. 3, 2: 402–410.
1892 c The earthworms of Middlesex. –Sci. Gossip 28: 194–196.
1892 d Studies of British tree- and earthworms. –J. Linn. Soc. Lond. (Zool.) 24: 292.
1893 A check-list of British earth-worms. –Naturalist, 1893: 17–20.
1897 Earthworm studies. IV. A check-list of British earthworms. –Zoologist, ser. 4, 1: 453–459.
1910 New garden worms. –Gardener's Chron., ser. 3, 48: 98–99.
1911 The distribution of British annelids. –Zoologist, ser. 4, 15: 143–146, 184–191, 367[374.
1921 Two new aquatic annelids. –Ann. Mag. Nat. Hist., ser. 9, 7: 137–141.

GANSEN, P. VAN
1963 Structures et fonctions du tube digestif du Lombricien *Eisenia foetida* Savigny –Ann. Soc. Zool. Belg. 93: 1–120.

GARMAN, H.
1888 On the anatomy and histology of a new earthworm (*Diplocardia communis* gen. et sp. nov.). –Bull. Ill. St. Lab. Nat. Hist. 3: 47–77.

GATES, G. E.
1941 Notes on a Californian earthworm, *Plutellus papillifer* (Eisen, 1893). –Proc. Calif. Acad. Sci. ser. 4, 23(4): 443–452.
1942 Check list and bibliography of North American earthworms. –Amer. Midl. Nat. 27(1): 86–108.
1943 On some American and Oriental earthworms. –Ohio J. Sci. 43(2): 87 98.
1949 Miscellanea megadrilogica. –Amer. Nat. 83(810): 139–152.
1952 New species of earthworms from the Arnold Arboretum, Boston. –Breviora, Mus. Comp. Zool., no. 9, 3 pp.

1953a On the earthworms of the Arnold Arboretum, Boston. –Bull. Mus. Comp. Zool., Harv., 107(10): 501–534.
1953b Further notes on the earthworms of the Arnold Arboretum, Boston. –Breviora, Mus. Comp. Zool., no. 15, 9 pp.
1954 Exotic earthworms of the United States. –Bull. Mus. Comp. Zool., Harv., 111(6): 219–258.
1956 Notes on American earthworms of the family Lumbricidae. III-VII. –Bull. Mus. Comp. Zool., Harv., 115(1): 1–46.
1958 Contribution to a revision of the earthworm family Lumbricidae. II. Indian species. –Breviora, Mus. Comp. Zool., no. 91, 16 pp.
1959 Earthworms of North American caves. –Natn. Speleol. Soc. Bull. 21(2): 77–84.
1961 Ecology of some earthworms with special reference to seasonal activity. –Amer. Midl. Nat. 66: 61–86.
1965 Louisiana earthworms. I. A preliminary survey. –Proc. La. Acad. Sci. 28: 12–20.
1966 Requiem—for megadrile utopias. A contribution toward the understanding of the earthworm fauna of North America. –Proc. Biol. Soc. Wash. 79: 239–254.
1967 On the earthworm fauna of the Great American desert and adjacent areas. –Gt. Basin Nat. 27(3): 142–176.
1968 What is *Enterion ictericum* Savigny 1826 (Lumbricidae, Oligochaeta)? –Bull. Soc. Linn. Normande, ser. 10, 9: 199–208.
1969 On two American genera of the earthworm family Lumbricidae. –J. Nat. Hist., Lond., 9: 305–307.
1970 Miscellanea megadrilogica VIII. –Megadrilogica 1(2): 1–14.
1972a Toward a revision of the earthworm family Lumbricidae. IV. The trapezoides species group. –Bull. Tall Timbers Res. Stn., no. 12, 146 pp.
1972b On American earthworm genera. I. *Eisenoides* (Lumbricidae). –Bull. Tall Timbers Res. Stn., no. 13: 1–17.
1972c Burmese earthworms. An introduction to the systematics of megadrile oligochaetes with special references to southeast Asia. –Trans. Amer. Philos. Soc. 62(7): 1–326.
1973a The earthworm genus *Octolasion* in America. –Bull. Tall Timbers Res. Stn., no. 14: 29–50.
1973b Contribution on the species name *Lumbricus terrestris*. –Bull. Zool. Nomencl. 30(2): 34.
1974a On oligochaete gonads. –Megadrilogica 1(9): 1–4.
1974b Contribution to a revision of the Lumbricidae. X. *Dendrobaena octaedra* (Savigny) 1826, with special references to the importance of its parthenogenetic polymorphism for the classification of earthworms. –Bull. Tall Timbers Res. Stn., no. 15: 15–57.
1974c Contributions to a revision of the family Lumbricidae. XI. *Eisenia rosea* (Savigny, 1826). –Bull. Tall Timbers Res. Stn., no. 16: 9–30.
1974d On a new species of earthworm in a southern portion of the United States. –Bull. Tall Timbers Res. Stn., no. 15: 1–13.
1975a Contributions to a revision of the earthworm family Lumbricidae. XII. *Enterion mammale* Savigny, 1826 and its position in the family. –Megadrilogica 2(1): 1–5.
1975b Contributions to a revision of the earthworm family Lumbricidae. XVII. *Allolobophora minuscula* Rosa, 1906 and *Enterion pygmaeum* Savigny, 1826. –Megadrilogica 2(6): 7–8.
1976 Contributions to a revision of the earthworm family Lumbricidae. XIX. On the genus of the earthworm *Enterion roseum* Savigny, 1826. –Megadrilogica 2(12): 4.
1977 More on the earthworm genus *Diplocardia*. –Megadrilogica 3(1): 1–47.

GAVRILOV, K.
1935 Contribution à l'étude de l'autofecondation chez les Oligochètes. –Acta Zool. Stockholm 16: 111–115.
1939 Sur la reproduction de *Eiseniella tetraedra* (Sav.) forma *typica*. –Acta Zool. Stockholm 20: 439–464.

GERARD, B. ..
1964 Synopses of the British fauna. no. 6.—Lumbricidae (Annelida) with keys and descriptions. –Linn. Soc. Lond., 58 pp.

GIAVANNOLI, L.
1933 Cave life of Kentucky. VII. Invertebrate life of Mammoth and other neighboring caves. –Amer. Midl. Nat. 14(5): 622–623.

GISH, C. D. and R. E. CHRISTENSEN
1973 Cadmium, nickel, lead and zinc in earthworms from roadside soil. –Environ. Sci. Tech. 7: 1060–1062.

GOFF, C. C.
1952 Flood-plain animal communities. –Amer. Midl. Nat. 47(2): 478–486.

GRAFF, O.
1953 Zur Berechtigung des Artnames *Lumbricus terrestris* Linnaeus, 1758. –Zool. Anz. 161(11/12): 324–326.

GRUBE, A. E.
1851 Annulaten. *In* Middendorff, A.T., Reise in den äussersten Norden und Osten Sibiriens. 2(1): 1–24.

GUILD, W. J. MCL.
1951 The distribution and population density of earthworms (Lumbricidae) in Scottish pasture fields. –J. Anim. Ecol. 20(1): 88–97.

HAGUE, F. S.
1923 Studies on *Sparganophilus eiseni* Smith. –Trans. Amer. Micros. Soc. 42(1): 1–38.

HARMAN, W. J.
1952 A taxonomic survey of the earthworms of Lincoln Parish, Louisiana. –Proc. La. Acad. Sci. 15: 19–23.
1954 Some earthworms from southern Oklahoma. –Proc. Okla. Acad. Sci. 35: 51–55.
1955 Earthworms of commercial importance and their effect on distribution. –Proc. La. Acad. Sci. 18: 54–57.
1960 Studies on the taxonomy and musculature of the earthworms of central Illinois. –Ph.D. thesis, Univ. of Illinois, 107 pp.
1965 Life history studies of the earthworm *Sparganophilus eiseni* in Louisiana. –Southwestern Nat. 10(1): 22–24.

HEIMBURGER, H. V.
1915 Notes on Indiana earthworms. –Proc. Ind. Acad. Sci. 25: 281–285.

HOFFMEISTER, W. F. L.
1842 De vermibus quibusdam ad genus Lumbricorum pertinentibus. –Dissertation, Berlin, 28 pp.
1843 Beiträge zur Kenntnis deutscher Landanneliden. –Arch. Naturg. 9(1): 183–198.
1845 Die bis jetzt bekannten Arten aus der Familie der Regenwürmer. Als Grundlage zu einer Monographie dieser Familie. –Branschweig, Vieweg, 43 pp.

HOPKINS, A. R. and V. M. KIRK
1957 Effects of several insecticides on the English red worm. –J. Econ. Ent. 50(5): 699–700.

HUNT, L. B. and R. J. SACHO
1969 Response of robins to DDT and methoxychlor. –J. Wildl. Manage. 33: 267–272.

HUTTON, F. W.
1877 On New Zealand earthworms in the Otago Museum. –Trans. N. Z. Inst. 9: 350–353.
1883 Synopsis of the genera of earthworms. –N. Z. J. Sci. 1: 585–586.

JANDA, V. and K. GAVRILOV
1939 Untersuchungen über die Vermehrungsfähigkeit von Individuen einiger Oligochaeten-Arten, die schon vor Erreichung der Geschlechtsreife isoliert wurden. –Vestnik Czech. Zool. Spolecuosti v Prag. 6–7: 254–259.

JEFFERSON, P.
1955 Studies on the earthworms of turf. A. The earthworms of experimental turf plots. –J. Sports Turf Res. Inst. 9(31): 6–27.

JOHNSTON, G.
1865 A catalogue of the British non-parasitical worms in the collection of the British Museum. –London, Taylor and Francis, 365 pp + 20 pls.

JOHNSTONE-WALLACE, D. B.
1937 The influence of wild white clover on the seasonal production and chemical composition of pasture herbage and upon soil temperatures. Soil moistures and erosion control. –4th Int. Grassl. Congr. Rep., pp. 188–196.

JORDAN, G. A., J. W. REYNOLDS and A. J. BURNETT
1976 Computer plotting and analysis of earthworm population distribution in Prince Edward Island. –Megadrilogica 2(10): 1–7.

JOYNER, J. W.
1960 Earthworms of the Upper Whitewater Valley (East-Central) Indiana. –Proc. Ind. Acad. Sci. 69: 313–319.

JUDD, W. W.
1964 Studies of the Byron Bog in southwestern Ontario. XIX. Distribution of earthworms. –Nat. Hist. Pap., Natn. Mus. Can., no. 25, 3 pp.
1970 Invertebrates from an abandoned muskrat house at London, Ontario. –Can. J. Zool. 48(2): 402–403.

JULIN, E.
1949 De Svenska daggmaskarterna. –Ark. Zool. 42A (17): 1–58.

KEVAN, D. K. MCE.
1962 Soil animals. –New York, Philosophical Library, 237 pp.

KINBERG, J. G. H.
1867 Annulata nova. Öfv. Vet.-Akad. Förh. Stockholm 23: 98.

KOBAYASHI, S.
1940 Terrestrial Oligochaeta from Manchoukou. –Sci. Rep. Tôhoku Imp. Univ., ser. 4, 15: 261–315.

LADELL, W. R. S.
1936 A new apparatus for separating insects and other arthropods from the soil. –Ann. Appl. Biol. 4: 862–879.

LANGMAID, K. K.
1964 Some effects of earthworm invasion in virgin podzols. –Can. J. Soil Sci. 44: 34–37.

LEGG, D. C.
1968 Comparison of various worm-killing chemicals. –J. Sports Turf Res. Inst. 44: 47–48.

LEUCKART, R.
1849 Zur Kenntnis der Fauna von Island. –Arch. Naturg. 15(1): 149–208.

LEVINSEN, G. M. R.
1884 Systematisk-geografisk oversigt over de nordiske Annulata, Gephyrea, Chaetognathi og Balanoglossi. –Vidensk. Meddel. Naturhist. Förh. Kjøbenhavn, ser. 4, 5: 92–350.

LINNAEUS, C.
1758 Systema naturae per regna tria naturae, secundum classes, ordines, genera, species cum characteribus, differentiis, synonymis, locis, Edition decima, reformata. Tom I. –Laurentii Salvii, Holmiae, 824 pp.

LIPA, J. J.
1958 Effect on earthworm and Diptera populations of BHC dust applied to soil. –Nature, Lond. 181: 863.

LISCINSKY, S.
1965 The American Woodcock in Pennsylvania. –Pennsylvania Game Comm., Harrisburg, 32 pp.

LJUNGSTRÖM, P. O.
1970 Introduction to the study of earthworm taxonomy. –Pedobiologia 10: 265–285.

LJUNGSTRÖM, P. O. and A. J. REINECKE
1969 Variation in the morphology of *Microchaetus modestus* (Microchaetidae: Oligochaeta) from South Africa with notes on its biology. –Zool. Anz. 182: 216–224.

LOGIER, E. B. S.
1958 The Snakes of Ontario. –Toronto, University of Toronto Press, x + 94 pp.

LOW, A. J.
1955 Improvements in the structural state of soils under leys. –J. Soil Sci. 6: 179–199.

MACNAB, J. A. and D. MCKEY-FENDER
1947 An introduction to Oregon earthworms with additions to the Washington list. –Northwest Sci. 21(2): 69–75.

MALEVIČ, I. I.
1949 Materialy k poznaniju doždevych červej Orechovo-Plodovych lesov južnoj Kirgizii. –Dokl. Akad. Nauk. SSSR (Biol.) 47: 397–400.

MALM, A. W.
1877 Om daggmasker, Lumbricina. –Öfv. Salsk. Hortik. Förh. Göteborg 1: 34–47.

MCLEOD, J. H.
1954 Note on a Staphylinid (Coleoptera) predator of earthworms. –Can. Ent. 86(5): 236.

MCMAHAN, M. L.
1976 Preliminary notes on a new megadrile species, genus, and family from the southeastern United States. –Medagrilogica 2(11): 6–8.

MICHAELSEN, W.
1889 Oligochäten des Naturhistorischen Museums in Hamburg. III. –Mitt. Mus. Hamburg 7(3): 1–12.
1890 Die Lumbriciden Mecklenburgs. –Arch. Verh. Freunde. Naturg. Mecklenb. 44: 48–54.
1891 Die Terricolen-Faune der Azoren. –Abh. Nat. Verh. Hamburg 11(2): 1–8.
1892 Terricolen der Berliner Zoologischen Sammlung II. –Arch. Naturg. 58(1): 209–261.
1900 a Die Lumbriciden-Fauna Nordamerikas. –Abh. Nat. Verh. Hamburg 16(1): 1–16.
1900 b Oligochaeta. Lief. 10. *In* Das Tierreich. –Berlin, Friedländer, xxix + 575 pp.
1903 a Die geographische Verbreitung der Oligochäten. –Berlin, Friedländer, vi + 186 pp.
1903 b Neue Oligochäten und neue Fundorte altbekannter. –Mitt. Naturh. Mus. Hamburg 19: 1–54.
1907 Neue Oligochäten von Vorder-Indien, Ceylon, Birma und den Andaman-Inseln. –Mitt. Naturh. Mus. Hamburg. 24: 143–188.
1908 Annelida. A. Oligochäten aus dem westlichen Kapland. –Denskschr. Med.-Naturw. Ges. Jena 13: 29–42.
1910 Zur Kenntnis der Lumbriciden und ihrer Verbreitung. –Ann. Mus. Zool. Acad. Sci. St. Petersburg 15: 1–74.
1913 Die Oligochäten des Kaplandes. –Zool. Jb. Syst. 34: 473–555.
1921 Zur Stammesgeschichte und Systematik der Oligochäten, insbesondere der Lumbriculiden. –Arch. Naturg. 86 (A-8): 130–141.

MICKEL, C. E.
1925 Notes on *Zygocystis cometa* Stein, a gregarine parasite of earthworms. –J. Parasit. 11: 135–139.

MOON, H. P.
1934 An investigation of the littoral region of Windermere. –J. Anim. Ecol. 3: 8–28.

MOORE, H. F.
1893 Preliminary account of a new genus of Oligochaeta. –Zool. Anz. 16: 333.
1895 On the structure of *Bimastos palustris* a new oligochaete. –J. Morph. 10(2): 473–496.

MOORE, J. P.
1906 Hirudinae and Oligochaeta collected in the Great Lakes region. –Bull. Bur. Fish. 25: 153–171.

MORGAN, C.
1970 Profitable earth worm farming. 12th ed. –Elgin, Shields Publ., 93 pp.

MORRIS, H. M.
1922 On a method of separating insects and other arthropods from the soil. –Bull. Ent. Res. 13: 197.

MORRISON, F. O.
1950 The toxicity of BHC to certain microorganisms. Earthworms and arthropods. –Ont. Ent. Soc. Ann. Rep. 80: 50–57.

MULDAL, S.
1952 The chromosomes of the earthworms. I. The evolution of polyploidy. –Heredity 6: 55–76.

MÜLLER, O. F.
1774 Vermium terrestrium et fluviatilium, seu animalium infusoriorum, helminthicorum, et testaceorum, non marionorum, succincta historia. Pars altera. Helminthica. –Havniae et Lipsiae, 1(2): 1–72.

MUNSELL COLOR COMPANY, INC.
1954 Munsell soil color charts. –Baltimore, i + 14 pp.

MURCHIE, W. R.
1956 Survey of the Michigan earthworm fauna. –Pap. Mich. Acad. Sci. Arts Lett. 41: 53–72.

MYERS, R.
1969 The ABC's of the earthworm business. –Elgin, Shields Publ., 60 pp.

NELSON, J. M. and J. E. SATCHELL
1962 The extraction of Lumbricidae from the soil with special reference to the hand sorting method. *In* Murphy, P. W., ed. Progress in Soil Zoology. –London, Butterworths, pp. 294–299.

OLIVER, J. H. JR.
1962 A mite parasitic in the cocoons of earthworms. –J. Parasit. 48(1): 120–123.

OLSON, H. W.
1928 The earthworms of Ohio, with a study of their distribution in relation to hydrogen-ion concentration, moisture and organic content of the soil. –Bull. Ohio Biol. Surv. 4(2), Bull. 17: 47–90.
1932 Two new species of earthworms for Ohio. –Ohio J. Sci. 32: 192–193.
1936 Earthworms of Missouri. –Ohio J. Sci. 36(2): 102–113.
1940 Earthworms of New York state. –Amer. Mus. Nov., no. 1090, 9 pp.

OMODEO, P.
1955a Lumbricidae e Lumbriculidae della Groenlandia. –Mem. Soc. Toscana Sci. Nat. Pisa, ser. B. 62: 105–128.
1955b Cariologia dei Lumbricidae. II Contributo. –Caryologia 8: 137–178.
1956 Contributo alla revisione dei Lumbricidae. –Arch. Zool. It. 41: 129–212.
1962 Oligochètes des Alpes. I. –Mem. Mus. Civ. Sto. Nat. Verona 10: 71–96.
1963 Distribution of the terricolous oligochaetes on the two shores of the Atlantic. *In* A. Löve and A. Löve (eds.). North Atlantic Biota and their history. Pp. 127–151.

ÖRLEY, L.
1881a A Magyarorszagi Oligochaetak faunaja. I. Terricolae. –Math. Term. Közlem. Magyar Akad. 16: 563–611.
1881b Beiträge zur Lumbricinen-Fauna der Balearen. –Zool. Anz. 4: 284–287.
1885 A palearktikus ovben elo terrikolaknak revizioja es elterjedese. –Ertek. Term. Magyar Akad. 15(18): 1–34.

PEARSE, A. S.
1946 Observations on the microfauna of the Duke Forest. –Ecol. Monogr. 16: 127–150.

PEECH, M.
1965 Exchange acidity, hydrogen-ion activity and lime requirement. *In* Black, C. A., ed. Me-

thods of Soil Analysis. Part 2, Chemical and Microbiological Properties. –Madison, Wisconsin, no. 9, pp. 905–932.

PERRIER, E.
1872 Récherche pour servir à l'historie des Lombriciens terrestres. –Nouv. Arch. Mus. Paris 8: 5–198.

PETTERSSEN, S.
1968 Introduction to meteorology. 3rd ed. –Toronto, McGraw-Hill, xi + 333 pp.

PICKFORD, G. E.
1926 On a new species of earthworm belonging to the subgenus *Bimastus* from Wicken Fen. –Ann. Mag. Nat. Hist., ser. 9, 17: 96–98.

PLISKO, J. D.
1961 Analiza materialów dżdżownic (Lumbricidae) zmagazynowanych prezez kreta (*Talpa europaea* L.). –Fragm. Faun. 9(7): 61–73.
1973 Lumbricidae dżdżownice (Annelida: Oligochaeta). –Fauna Polski, no. 1, Warszawa, 156 pp.

POP, V.
1941 Zur Phylogenie und Systematik der Lumbriciden. –Zool. Jb. Syst. 74: 487–522.
1948 Lumbricidele din România. –Ann. Acad. Repub. Pop. Române, Sect. Ştiint. Geol. Geogr. Biol., ser. A, 1(9): 383–507.

RAW, F.
1959 Estimating earthworm populations by using formalin. –Nature, Lond. 184: 1661–1662.
1960 Earthworm population studies: a comparison of sampling methods. –Nature, Lond. 187: 257.

RAW, F. and J. R. LOFTY
1959 Earthworm populations in orchards. –Rep. Rothamstead Exp. Stn. 1958: 134–135.

REDDELL, J. R.
1965 A check list of the cave fauna of Texas. I. The invertebrata (excl. of Insecta). –Texas J. Sci. 17: 143–187.

REINECKE, A. J. and P. A. J. RYKE
1969 A new species of the genus *Geogenia* (Microchaetidae: Oligochaeta) from Lesotho, with notes on two exotic earthworms. –Rev. Écol. Biol. Sol 6(4): 515–523.

REYNOLDS, J. W.
1972a Earthworms (Lumbricidae) of the Haliburton Highlands, Ontario, Canada. –Megadrilogica 1(3): 1–11.
1972b The activity and distribution of earthworms in tulip poplar stands in the Great Smoky Mountains National Park, Sevier County, Tennessee (Acanthodrilidae, Lumbricidae and Megascolecidae). –Cont. N. Amer. Earthworms (Annelida), no. 2, Bull. Tall Timbers Res. Stn., no. 11, pp. 41–54.
1972c A contribution to the earthworm fauna of Montana. –Proc. Mont. Acad. Sci. 32: 6–13.
1972d Répartition et biomasse des Lombricidés (Oligochaeta: Lumbricidae) dans trois bois de la France (Département d'Essonne). –Megadrilogica 1(4): 1–6.
1972e The relationship of earthworm (Oligochaeta: Acanthodrilidae and Lumbricidae) distribution and biomass in six heterogeneous woodlot sites in Tippecanoe County, Indiana. –J. Tenn. Acad. Sci. 47(2): 63–67.
1973a The earthworms of Delaware (Oligochaeta: Acanthodrilidae and Lumbricidae). –Megadrilogica 1(5): 1–4.
1973b The earthworms of Rhode Island (Oligochaeta: Lumbricidae). –Megadrilogica 1(6): 1–4.
1973c The earthworms of Connecticut (Oligochaeta: Lumbricidae, Megascolecidae and Sparganophilidae). –Megadrilogica 1(7): 1–6.
1973d Earthworm (Annelida, Oligochaeta) ecology and systematics. *In* Dindal, D. L., ed. Proc. 1st Soil Microcommunities Conf. –Springfield, U.S. Atomic Energy Commn., Natn. Tech. Inform. Serv., U.S. Dept. Com., pp. 95–120.

1973 e Review of Gates, G. E.: Burmese earthworms—An introduction to the systematics and biology of megadrile oligochaetes with special references to southeast Asia. –Syst. Zool. 22(2): 197–199.

1974 a Checklist, distribution and key to the Lumbricidae in Tennessee. –J. Tenn. Acad. Sci. 49(1): 16–20.

1974 b The earthworms of Maryland (Oligochaeta: Acanthodrilidae, Lumbricidae, Megascolecidae and Sparganophilidae). –Megadrilogica 1(11): 1–12.

1974 c Are oligochaetes really hermaphroditic amphimictic organisms? –Biologist 56(2): 90–99.

1975 a Boiteagan (Oligochaeta: Lumbricidae) Cheap Breatunn. –Megadrilogica 2(6): 1–7.

1975 b Les Lombricidés (Oligochaeta) des Îles-de-la-Madeleine. –Megadrilogica 2(3): 1–8.

1975 c The earthworms of Prince Edward Island (Oligochaeta: Lumbricidae). –Megadrilogica 2(7): 4–10.

1975 d Les Lombricidés (Oligochaeta) de l'Île d'Orléans, Québec. –Megadrilogica 2(5): 8–11.

1975 e Les Lombricidés (Oligochaeta) de la Gaspésie, Québec. –Megadrilogica 2(4): 4–9.

1975 f Die Biogeografie van Noorde-Amerikaanse Erdwurms (Oligochaeta) Noorde van Meksiko. I. –Indikator 7(4): 11–20.

1976 a The distribution and ecology of the earthworms of Nova Scotia. –Megadrilogica 2(8): 1–7.

1976 b Un aperçu des vers de terre dans les forêts nord-américaines, leurs activités et leurs répartition. –Megadrilogica 2(9): 1–11.

1976 c Catalogue et clé d'identification des lombricidés du Québec. –Nat. can. 103(1): 21–27.

1976 d A preliminary checklist and distribution of the earthworms of New Brunswick. –New Brunswick Nat. 7(2): 16–17.

1976 e *Aporrectodea icterica* (Savigny, 1826) une espèce de vers de terre récemment découverte en Amérique du Nord. –Megadrilogica 2(12): 3–4.

1976 f Die Biogeografie van Noorde-Amerikaanse Erdwurms (Oligochaeta) Noorde van Meksiko. II. –Indikator 8(1): 6–20.

1977 The earthworms of Massachusetts (Oligochaeta: Lumbricidae, Megascolecidae and Sparganophilidae). –Megadrilogica 3(2): 49–54.

In prep. The earthworms of Tennessee (Oligochaeta). II. Sparganophilidae.

In prep. Earthworms preservation for maximum use.

REYNOLDS, J. W. and D. G. COOK

1977 Nomenclatura Oligochaetologica, a catalogue of names, descriptions and type specimens of the Oligochaeta. –Fredericton, The University of New Brunswick. x + 217 pp.

REYNOLDS, J. W. and G. A. JORDAN

1975 A preliminary conceptual model of megadrile activity and abundance in the Haliburton Highlands. –Megadrilogica 2(2): 1–9.

REYNOLDS, J. W. and W. M. REYNOLDS

1972 Earthworms in medicine. –Amer. J. Nurs. 72(7): 1273.

REYNOLDS, J. W., E. E. C. CLEBSCH and W. W. REYNOLDS

1974 The earthworms of Tennessee (Oligochaeta). I. Lumbricidae. –Cont. N. Amer. Earthworms (Oligochaeta), no. 13, Bull. Tall Timbers Res. Stn., no. 17, vii + 133 pp.

REYNOLDSON, T. B.

1955 Observations on the earthworms of North Wales. –North Western Naturalist 3(3): 291–304.

RHEE, J. A. VAN

1969 Inoculation of earthworms in a newly drained polder. –Pedobiologia 9: 128–132.

RIBAUCOURT, E. DE

1896 Étude sur la faune lombricide de la Suisse. –Rev. Suisse Zool. 4: 1–110.

RIBAUCOURT, E. DE and A. COMBAULT

1906 Utilité des vers de terre en agriculture. –Rev. Gen. Agron. 1: 374–384.

RISSO, A.

1826 Histoire naturelle des principales productions de l'Europe méridionale et particulièrement de celles des environs de Nice et des Alpes maritimes. –Paris, Levrault, vol. 4, vii + 439 pp + 12 pls. (Les Lombrics, pp. 426–427).

ROOTS, B. I.
1956 The water relations of earthworms II. Resistance to desiccation and immersion, and behaviour when submerged and when allowed a choice of environment, –J. Exp. Biol. 33: 29–44.

ROSA, D.
1882 Descrizione di due nuovi Lumbrici. –Atti Acc. Torino 18: 169–173.
1884 Lumbricidi del Piemonte. –Torino 1884, pp. 5–54.
1886 Note sui lombrici del Veneto. –Atti Inst. Veneto, ser. 6, 4: 673–687.
1887 La distribuzione verticale dei lombrichi sulle Atpi. –Boll. Mus. Zool., Torino 2(31): 1–3.
1893 Revisione dei Lumbricidi. –Mem. Acc. Torino, ser. 2, 43: 399–476.

ROWE, J. S.
1972 Forest regions of Canada. –Can. For. Serv., Publ. 1300, x + 172 pp. + 2 pls.

SALT, G. and F. S. J. HOLLICK
1944 Studies of wireworm populations. I. A census of wireworms in pasture. –Ann. Appl. Biol. 81: 52–64.

SATCHELL, J. E.
1955 Some aspects of earthworm ecology. *In* Kevan, D. McK., ed. Soil zoology. –London, Butterworths, pp. 180–203.
1961 An electrical method of sampling earthworm populations. *In* Rodale, R. ed. The challenge of earthworm research. –Emmas, Soil and Health Foundation, pp. 89–97.
1967 Lumbricidae, *In* Burges, A. and F. Raw, eds. Soil biology. –New York, Academic Press, pp. 259–322.
1969 Methods of sampling earthworm populations. –Pedobiologia 9: 20–25.

SAVIGNY, J. C.
1826 Analyses des travaux de l'Academie Royale des Sciences pendant l'annee 1821, partie physique. Cuvier, M. le Baron G., ed. –Mém. Acad. Sci. Inst. Fr. 5: 176–184.

SAWYER, J. O. and A. A. LINDSAY
1964 The Holdridge bioclimatic formations of the eastern and central United States. –Proc. Ind. Acad. Sci. 72: 105–112.

SCHREAD, J. C.
1952 Habits and control of the oriental earthworm. –Conn. Agric. Exp. Stn. 556: 5–15.

SCHWERT, D. P.
1976 Recent records of earthworms (Oligochaeta: Lumbricidae) from central New York state. –Megadrilogica 2(10): 7–8.
1977 The first North American record of *Aporrectodea icterica* (Savigny, 1826) (Oligochaeta, Lumbricidae), with observations on the colonization of exotic earthworm species in Canada. –Can. J. Zool., 55(1): 245–248.

SHIELDS, E. B.
1971 Raising earthworms for profit. 9th ed. –Elgin, Shields Publ.

SIMS, R. W.
1973 *Lumbricus terrestris* Linnaeus, 1758 (Annelida, Oligochaeta): designation of a neotype in accordance with accustomed usage. Problems arising from the misidentification of the species by Savigny (1822 and 1826). –Bull. Zool. Nomencl. 30(1): 27–33.

SKOCZEŃ, S.
1970 Gromadzenie zapasów pokarmowych przez neiktóre ssaki owadożerne (Insectivora). –Przeglad Zool. 14(2): 243–248.

SMITH, F.
1895 A preliminary account of two new Oligochaeta from Illinois. –Bull. Ill. St. Lab. Nat. Hist. 4(5): 142–147.
1896 Notes on species of North American Oligochaeta II. –Bull. Ill. St. Lab. Nat. Hist. 4(14): 396–413.

1915 Two new varieties of earthworms with a key to the described species in Illinois. –Bull. Ill. St. Lab. Nat. Hist. 10(8): 551–559.
1917 North American earthworms of the family Lumbricidae in the collections of the United States National Museum. –Proc. U.S. Natn. Mus. 52(2174): 157–182.
1928 An account of changes in the earthworm fauna of Illinois and a description of one new species. –Ill. Div. Nat. Hist. Surv. 18(10): 347–362.

SMITH, F. and B. R. GREEN
1916 The Porifera, Oligochaeta, and certain other groups of invertebrates in the vacinity of Douglas Lake, Michigan. –Mich. Acad. Sci., Ann. Rep., 17: 81–84.

SMITH, W. W.
1887 Notes on New Zealand earthworms. –Trans. N.Z. Inst. 19: 122–139.
1894 Further notes on New Zealand earthworms. –Trans. N.Z. Inst. 26: 155–175.

SOUTHERN, R.
1909 Contribution towards a monograph of the British and Irish Oligochaeta. –Proc. R. Irish Acad. 27B(8): 119–182.
1910 A new species of enchytraeid worm from the White Mountains. –Proc. Acad. Nat. Sci. 62: 18–20.

STAFFORD, J.
1902 Notes on worms. –Zool. Anz. 25: 481–483.

STEBBINGS, J. H.
1962 Endemic-exotic earthworm competition in the American midwest. –Nature, Lond. 196(4857): 905–906.

STEPHENSON, J.
1917 On a collection of Oligochaeta from various parts of India and further India. –Rec. Indian Mus. 13: 353–416.
1922 Some earthworms from Kashmir, Bombay and other parts of India. –Rec. Indian Mus. 22: 427–443.
1923 The fauna of British Indian, including Ceylon and Burma. Oligochaeta. –London, Taylor and Francis, xxiv + 518 pp.
1930 The Oligochaeta. –Oxford, Clarendon Press, xvi + 978 pp.

STONE, P. C. and D. G. OGLES
1953 *Uropoda agitans*, a mite pest in commercial fishworm beds. –J. Econ. Ent. 46(4): 711.

STØP-BOWITZ, C.
1969 A contribution to our knowledge of the systematics and zoogeography of norwegian earthworms (Annelida Oligochaeta: Lumbricidae). –Nytt. Mag. Zool. 17(2): 169–280.

SVENDSEN, J. A.
1955 Earthworm population studies: a comparison of sampling methods. –Nature, Lond. 175: 864.
1957 The behaviour of lumbricids under moorland conditions. –J. Anim. Ecol. 26: 423–439.

SWARTZ, R. D.
1929 Modification of behavior in earthworms. –J. Comp. Psychol. 9: 17–33.

TANDY, R. E.
1969 The earthworm genus *Pheretima* Kinberg, 1866 in Louisiana. –Ph.D. thesis, Louisiana State Univ. 155 pp.

TAUBER, P.
1879 Annulata Danica. (1) En kritisk revision af de i Danmark fundene Annulata, Gephyrea, Balanoglossi, Discophorae, Oligochaeta, Gymnocopa og Polychaeta. –Kjøbenhavn. 144 pp.

TEMPLETON, R.
1836 A catalogue of the species of annulose animals. –Ann. Mag. Nat. Hist. 9: 233–240.

TÉTRY, A.
1934 Description d'une espèce française du genre *Pelodrilus*. –C. R. Acad. Sci. Paris 199: 322.

1936 Une variété nouvelle de *Lumbricus castaneus* Savigny, sa valeur systématique. -Bull. Soc. Sci. Nancy 1936, pp. 196-201.

1937 Révision des lombriciens de la collection de Savigny. Bull. Mus. Hist. Nat. Paris 9: 140-155.

1938 Contribution à l'étude de la faune de l'est de la France (Lorraine). -Nancy, Georges Thomas, 453 p.

THOMAS, M. K.

1953 Climatological atlas of Canada. -Natn. Res. Coun., Div. Bldg. Res. 3151(41): 253 pp.

THOMPSON, A. R.

1971 Effects of nine insecticides on the numbers and biomass of earthworms in pasture. -Bull. Environ. Contr. Toxicol. 5(6): 577-586.

THOMSON, A. J. and D. M. DAVIES

1973a The biology of *Pollenia rudis*, the cluster fly (Diptera: Calliphoridae). I. Host location by first-instar larvae. -Can. Ent. 105: 335-341.

1973b The biology of *Pollenia rudis*, the cluster cly (Diptera: Calliphoridae). II. Larval feeding behaviour and host specificity. -Can. Ent. 105: 985-990.

1974 Production of surface casts by the earthworm *Eisenia rosea*. -Can. J. Zool. 52(5): 659.

THORNTHWAITE, C. W.

1948 An approach toward a rational classification of climate. -Geogr. Rev. 38(1): 55-94.

TUZET, O.

1946 Sur la spermatogenèse atypique des lombriciens. -Arch. Zool. Exp. Gen. 84: 155-168.

UDE, H.

1885 Über die Rückenporen der Terricolen Oligochäten, nebst Beiträgen zur Histologie des Lieberaschlauches und zur Systematik der Lumbriciden. -Z. Wiss. Zool. 43: 87-143.

1905 Terricole Oligochäten von der Inseln der Südsee und von verschiedenen andern Gebieten der Erde. -Z. Wiss. Zool. 83: 405-501.

U.S. DEPT. AGRIC. AGRICULTURAL RESEARCH ADMINISTRATION

1951 Soil survey mannual. -Washington, U.S. Gov. Printing Off. U.S. Dept. Agric. Handb. no. 18, vii + 503 pp.

U.S. DEPT. AGRIC. SOIL SURVEY STAFF. SOIL CONSERV. SERV.

1964 Soil Classification, a comprehensive system (7th approximation). -Washington, U.S. Gov. Printing Off., 265 pp.

1967 Supplement to soil classification system (7th approximation). -Washington, U.S. Gov. Printing Off., 207 pp.

1968 Supplement to soil classification system (7th approximation), Histosols. -Washington, U.S. Gov. Printing Off., 22 pp.

VAIL, V. A.

1972 Natural history and reproduction in *Diplocardia mississippiensis* (Oligochaeta). -Cont. N. Amer. Earthworms (Annelida), no. 1, Bull. Tall Timbers Res. Stn., no. 11, pp. 1-39.

1974 Observations on the hatchlings of *Eisenia foetida* and *Bimastos tumidus* (Oligochaeta: Lumbricidae). -Cont. N. Amer. Earthworms (Annelida), no. 11, Bull. Tall Timbers Res. Stn., no. 16, pp. 1-8.

VAILLANT, L.

1889 Histoire naturelle des Anneles marins et d'eau douce. -Hist. Nat. Annel. 3(1): 1-113.

VÉDOVINI, A.

1967 Une nouvelle forme provençale d'*Allolobophora rosea* (Savigny). -Bull. Soc. Zool. Fr. 92: 793-796.

VEJDOVSKÝ, F.

1875 Beiträge zur Oligochaetenfauna Böhmens. Sitzber. Böhm. Ges. Wiss. Prag., pp. 191-201.

1882 Thierische Organismen der Brunnenwasser von Prague. -Prague.

1884 System und morpholgie der Oligochäten. -Prague, Rivnae. 136 pp.

1888 Die Entwicklungsgeschichte der Oligochäten, mit atlas. –Prag., Entwickgesch. Unters. 401 pp.

WALLWORK, J. A.

1970 Ecology of Soil Animals. –London, McGraw-Hill. 283 pp.

WALTON, W. R.

1933 The reaction of earthworms to alternating currents of electricity in the soil. –Proc. Ent. Soc. Wash. 35(2): 24–27.

WATERS, R. A. S.

1955 Numbers and weights of earthworms under a highly productive pasture. –N.Z.J. Sci. Tech., Sect. A, Agric. Res. 36: 516–525.

WHITEHOUSE, R. H. and A. J. GROVE

1943 The dissection of the earthworm. –London, Univ. Tutorial Press. 72 pp.

WICKETT, W. P.

1967 Exotic specimens of earthworms in British Columbia eaten by Coho salmon. –J. Fish. Bd. Can. 24(6): 1421–1422.

WILLIAMS, G. W.

1942 Observations on several species of *Metaradiophrya* (Protozoa: Giliata). –J. Morph. 70: 545–572.

YAHNKE, W. and J. A. GEORGE

1972 Rearing and immature stages of the cluster fly (*Pollenia rudis*) (Diptera: Calliphoridae) in Ontario. –Can. Ent. 104: 567–576.

ZAJONC, I.

1970 Synúzie dážd̆oviek (Lumbricidae) na lúkach Karpatskej oblasti Československa. –Biol. Práce 16(8): 5–98.

ZICSI, A.

1959 Faunistisch-systematische und ökologische Studien über die Regenwürmer Ungarns. I. Acta Zool. –Hung. 5(1-2): 165–189.

1962 Über die Dominanzhältnisse einheimischer Lumbriciden auf Ackerboden. –Opusc. Zool. Budapest 4(2-4): 157–161.

Date D[illegible]